MIND IN **ARCHITECTURE**

MIND IN **ARCHITECTURE**

NEUROSCIENCE, EMBODIMENT, AND THE FUTURE OF DESIGN

edited by Sarah Robinson
and Juhani Pallasmaa

The MIT Press
Cambridge, Massachusetts
London, England

MIT Press books may be purchased at special quantity discounts for business or sales promotional use. For information, please email special_sales@mitpress.mit.edu.

This book was set in Frutiger and Sabon by the MIT Press. Printed and bound in the United States of America.

Library of Congress Cataloging-in-Publication Data
Minding Design (Symposium) (2012 : Scottsdale, Ariz.)
Mind in architecture : neuroscience, embodiment, and the future of design / edited by Sarah Robinson and Juhani Pallasmaa.
pages cm
"This book has its origins at the Minding Design symposium that took place at Taliesin West in November 2012, an event sponsored by the Frank Lloyd Wright Foundation and Taliesin, the Frank Lloyd Wright School of Architecture."
Includes bibliographical references and index.
ISBN 978-0-262-02887-5 (hardcover : alk. paper)
1. Neurosciences in architecture—Congresses. 2. Architecture—Human factors—Congresses. 3. Architectural design—Psychological aspects—Congresses. I. Robinson, Sarah (Architect), editor. II. Pallasmaa, Juhani, editor. III. Title.
NA2543.N48M56 2012
720.1'05—dc23

2014034234

10 9 8 7 6 5 4 3 2 1

CONTENTS

ACKNOWLEDGMENTS

To write and edit this book was to feel oneself part of an evolutionary process. The knowledge brought to us by the cognitive and neurosciences is developing so rapidly, and the implications of this research are so far-reaching, that one is nowadays in a constant state of revision. Nothing here would have been possible without the generosity and curiosity of our fine cast of contributors—it is to them that we owe our greatest debt. This book has its origins at the "Minding Design" symposium that took place at Taliesin West in November 2012, an event sponsored by Taliesin, the Frank Lloyd Wright School of Architecture. Taliesin has continued to support the evolution of the dialogue that was opened there by supporting the production of this book. We wish to thank them and to acknowledge Maura Grogan, Susan Jacobs Lockhart, David Mohney, and Victor Sidy, in particular, for championing our project when it was but a glimmer in the eye. Aris Georges's considerable talent for graphic design is evident on every page of this book. And Roger Conover and Justin Kehoe at the MIT Press have consistently proven to be perceptive, patient, professional, and caring. Gillian Beaumont edits with a keen eye and a fine sieve; her attention and insight made this a better book—one that would not have been possible without all of these collective efforts.

INTRODUCTION: SURVIVAL THROUGH DESIGN

Sarah Robinson

A growing number of us ... are convinced that generally valid scales and gauges for judging design ... can be found and must be applied. To deny it would seem nihilistic.[1]

Richard Neutra, 1954

Rain draws sap from creosote leaves, releasing their astringent odor into the air. Rain washes hard earth; sinks into the saguaro's radial roots—uncreasing the crinoline folds of its skin. Rain enervates the landscape; turns cactus needles into hairs that stand on end, antennae tuned to capture water. The raindrops beat on my metal roof like a drum. From below, veils of canvas draw the line thinly between outside and in. The south-facing wall arches against my back like the cupped palm of a giant hand, exhaling its stored heat into the length of my spine. The wind outside howls. The fire inside cracks and pops. During a storm we experience dwelling most intensely. "Faced with the bestial hostility of the storm ... the house's virtues of protection and resistance are transposed into human virtues. The house acquires the physical and moral energy of a human body ... it is an instrument with which to confront the cosmos," wrote Gaston Bachelard.[2]

Bachelard was not the first to compare the energy of the body to the agency of a building. Here the body is a metaphor in the fullest sense of the term; that is, a direct transfer of meaning from one domain to another, effectively linking two previously separate entities. In this case, the dwelling's latent capacity to protect is activated by the storm. The house shifts from a state of dormancy to one of active response. The storm charges the atmosphere; awakens the house, converting it into a sensitive instrument—a living container endowed with physical and moral energy. To truly appreciate the power of this dynamic, one must experience it firsthand. Witnessing the many moods of the Sonoran Desert from my shelter at Taliesin West afforded me that opportunity: total immersion in the didactics of body, dwelling, and desert. This is exactly what Frank Lloyd Wright intended when he sent his apprentices to live in tents and to build shelters in the desert. With a Zen master's methodological flair, he sought to create the conditions in which this visceral, omnisensorial flash of knowing could occur. He considered this experience to be the fundamental existential ground for becoming an architect. In order to design a building that is a breathing entity, not an inert box, one must first experience what it means to truly dwell. Wright understood that profound learning could not take place until the student sensed her own identity and belonging within this broader continuity of self, building, and world. Such an experiential foundation was also the very essence of John Dewey's educational philosophy. For Dewey, bodily experience was the primal ground for everything we think, know, mean, or communicate. Dewey's philosophy was not only pragmatic, it was empirically responsible and informed by the best physiology, psychology, and neuroscience of his time. When Wright founded his own school, he based its pedagogy on Dewey's core principles. The central tenet of the Taliesin Fellowship, "to learn by doing," was the distillation of Dewey's entire philosophy of experience. In the original 1932 prospectus in which Wright listed Dewey among the "Friends of Taliesin," students were offered an apprenticeship with a master. This was the classic technique for learning martial arts, painting, sculpture, woodworking, masonry, and violin making—arts and crafts traditions honored and practiced throughout the world for centuries. In this context learning occurred not through the study of texts so much as it was directly transferred from the embodied knowledge of the master to the receptive, experiential ground of the student.

Yet as a pedagogy for architecture, the apprenticeship model stood in sharp contrast to the conventional architectural education of the time that had long since abandoned embodied knowledge for purely intellectual pursuits. Wright and Dewey were among the few twentieth-century thinkers who understood the full richness, complexity, and philosophical importance of embodied experience. In fact, the primacy of embodied experience in architectural education could be considered one of Wright's largely neglected contributions to modern architecture.

For the past three decades, while architectural theory was following the other humanities into the vertiginous and barren peaks of semiotics, experts in numerous branches of the sciences were moving in exactly the opposite direction. Rather than getting caught up in convoluted cerebral games that denied emotional and bodily reality altogether, researchers were piling up evidence for the bodily basis of mind and meaning. Diverse disciplines such as biology, psychology, cognitive neuroscience, and phenomenology steadily yielded evidence of the extent to which mental properties depend on the functioning of the human nervous system. They collectively converged on this fact: all human endeavors depend upon our brains functioning as organic members of our bodies, which are in turn actively engaged with the ecological, architectural, social, and cultural environments in which we dwell. Embodiment calls for a far-reaching reconceptualization of who and what we are, in a way that contradicts much of our Western philosophical and religious heritage. To accept that our minds can include aspects of our physical and cultural environments means that the kinds of environments we create can alter our minds and our capacity for thought, emotion, and behavior. Such an assertion undermines the certainty of our ontological categories—the dichotomies that separate the inside from the outside and the subject from the object are not distinctions of kind as much as they are abstractions that arise from our ongoing interaction in the world. The notion of a separate self that operates in isolation from its environment is thus tossed into the wreckage of an outworn paradigm. We can trace Dewey's nondualistic thinking forward through recent cognitive science in the work of Varela and Maturana, Alva Noë and George Lakoff, in neuroscience most notably in the work of Gerald Edelman, Vittorio Gallese, and Giacomo Rizzolatti, through to the phenomenological investigations of Mark Johnson—all of whom argue that the reality of embodiment mandates a radical reevaluation of long-cherished dogmas that split the material from the intellectual, the mind from the body, and the self from the environment. While the implications of this provocative claim have not quite entered public consciousness, elsewhere entire research programs are being scrapped and redirected; new disciplines are emerging to confront this newly sanctioned reality. As Harry Mallgrave states in this book, "It is no exaggeration to say that we have learned more about our biological selves in the past half-century than in all of human history, and as a result of these developments the humanities—sociology, philosophy, psychology, and human paleontology in particular—have been forced to restructure their premises and research agendas radically. Yet architects have remained surprisingly incurious or seem little moved by these events." Like many other professions, the practice of architecture is in crisis, and though we now spend 90 percent of our time in buildings, architects design a mere fraction of them. Meanwhile, the cognitive and neurosciences, and the embodiment theory on which they are based, are revolutionizing knowledge across disciplines. In fact, some observers have suggested that we are in the midst of a revolution in neuroscience that is as significant as the Galilean revolution in physics and the Darwinian revolution in biology. A concert of experts spanning disciplines affirm the critical role that the environment—built and natural—plays in determining our mental, physical, cultural, and social evolution. As architects, we are not only aware of such a premise; many of us have staked our careers on it. While we may not need to be convinced of the decisive role that the built environment plays in directing human behavior and evolution, our clients, our heirs, and our public certainly do.

Understanding architecture's role in shaping who we are and what we might become stands to increase the gravity of our work, elevate the stature of our role, and highlight the extent of our contribution to human and ecological well-being. To ignore the potential impact that neuroscientific research has upon architectural education and practice is to miss an extraordinary opportunity, since we are the very group that this new knowledge could most persuasively serve.

Because architecture bridges both science and art, architects have long applied principles gleaned from scientific research into our practice. An early proponent of using neuroscience to inform architecture was Richard Neutra. Since he worked with Wright during the time that the original fellowship was being formed, it would be only a slight stretch to consider him to be a first-generation Taliesin apprentice. He wrote his book *Survival through Design* during the war years. While the book's title might sound extreme—the notion that human survival could somehow be secured through design—the promises of modernism were exactly that. Modernist architecture had the potential to liberate us from the bondage of cramped thinking undertaken in dark quarters: it would increase our efficiency, expand our mental and physical horizons, and guarantee a brighter future. Neutra championed these noble aims, and was convinced that design should be based on a biological understanding of human nature. "Our time is characterized by a systematic rise of the biological sciences and is turning away from oversimplified mechanistic views of the 18th and 19th centuries, without belittling in any way the temporary good such views may have once delivered. An important result of this new way of regarding this business of living may be to bare and raise appropriate working principles and criteria for design,"[3] he wrote.

Survival through Design is the fruit of a lifetime of design experience supplemented with his own research in psychology and physiology. The fact that he dedicated this book to Wright can be regarded as evidence of Neutra's allegiance to the lineage of embodied thinking and empirical investigation that Taliesin engendered. Neutra carried on the tradition to which Wright himself belonged—one that understood the whole person in the context of a broader ecology that included biological, social, cultural, and linguistic realms; one that was guided by principles won through "tangible observation rather than abstract speculation."[4] This embodied empiricism is consonant with the knowledge being advanced in the best of contemporary cognitive science, neuroscience, and phenomenology. Though this appreciation of our embodiment may be breaking new ground in other disciplines, architecture's earliest manifestations have always been rooted in corporeal experience. Vitruvius summoned a thousand years of classical culture in his formulation of the "well-shaped man," a legacy carried forward by Leon Battista Alberti, who considered the ideal building to be an emulation of the human body. This celebration of human embodiment was present through to the Renaissance, and only later was the human abstracted out of the architectural equation. With some notable exceptions—like the case I have tried to build for Taliesin—human bodily experience has been effectively eradicated from architectural education and practice right up until the present day. So, while the findings of neuroscience and the embodied empiricism in which they are rooted may not be novel within the panorama of architectural history, they can nevertheless serve as a catalyst for reinstating the humanity that we have lost along the way. As Neutra foreordained, biological science in general and neuroscience in particular are finally poised to provide some basic working principles and criteria for design. The chapters in this volume critically consider what some of these principles and criteria might be. Many of them have their origins at the symposium "Minding Design: Neuroscience, Design Education and the Imagination," a collaboration between the Frank Lloyd Wright School of Architecture and the Academy of Neuroscience for Architecture that brought leading neuroscientists and leading architects to the same table: a table that was set in the campus of Taliesin West.

As any skilled host or hostess knows, the success of the dinner party often hinges on the atmosphere of the setting. The setting not only shelters the table but influences the course, content, and mood of the conversations that happen around it. Wright's desert masterpiece is a case in point. The disciplines of neuroscience and architecture intersect in their understanding of and obligation to their subject: the embodied human being, a being who can exist only in relationship; relationship to the places we inhabit, to each other, to the world. Architectural settings have the capacity to foster, weaken, or destroy these relationships. Both the neuroscientists involved in the Academy of Neuroscience for Architecture who work at Louis Kahn's Salk Institute and the students who study at Taliesin are aware of this interdependence.

That weekend in the desert was a point of departure, a catalyst that cracked open a few new windows and opened others wide enough to knock the corner post out of the wall. Momentum has since gathered, and so have the contributions to this work. Here, practitioners from such disciplines as psychiatry, neuroscience, physiology, philosophy, cognitive science, architectural history, and architectural practice meet not only to explore what neuroscience and architecture can learn from each other: they situate the dialogue in historic context; they examine the implications for current practice and in reimagining architectural education; they dream the shape of the future. What emerge here are design criteria that have been forged over eons of evolution on this planet—whose imperatives are neither arbitrary nor negotiable. Attention is not narrowed in algorithms, signifiers, and particles, but directed toward the emergent, the affective, the sensual, the gestural and kinesthetic factors that pattern human perception and experience. Here you will find an increasingly complex, subtilized understanding of human drives, longings, and desires, and arguments that detail the unequivocal agency of the buildings we create. Longstanding concerns surface in new guises, old dichotomies are redressed, priorities realigned, once fashionable theories are shaken, former allegiances are questioned and new avenues entertained. In these chapters you will find neither dogma nor ideological certainty poised to fill the void left by the once cherished theories we are forced to discard. Instead, this is a beginning—a point within a larger process. Truly reckoning with the implications of embodiment and taking the findings yielded by the new sciences of the mind seriously is a vast undertaking—one that is inherently multidisciplinary and necessarily collaborative. One that calls for a "co-evolution of theories,"[5] as the philosopher Patricia Churchland has suggested. Evolution is a process that advances in fits and starts, bootstrapped to a plurality of conditions, impulses, drives—it works from the bottom up, *and* the top down.

Claims that "neuroscientists will be the next great architects"[6] betray a misunderstanding of the roles and intrinsically interdependent nature of both professions. Despite popular mythologizing to the contrary, architects are and always have been team players. We are more like film directors, dependent on the interactive chemistry between the location, the actors, the story, and the budget, than we are solitary dictators with transcendent visions. With refined sensibility, experience, and shared knowledge, we are able to fashion a whole from a host of diverse variables. A lesser building might be a kit of parts. Our best work emerges from a synthesis that transcends every one of its variables to become complete, a unique entity—a new world. To accomplish this, we must be generalists who listen to and depend upon the expertise of our collaborators. What we make matters. Our responsibility is ethically grounded in our human past, and only in honoring our bio-historical roots can we hope to design a sustainable future. "Today design may exert a far-reaching influence on the nervous make-up of generations,"[7] Neutra wrote six decades ago. Indeed, architecture is not optional—it is not a luxury item—it is and always has been the very fabric of our survival, our potential flourishing, or our possible demise.

NOTES

1. Richard Neutra, *Survival through Design* (Oxford: Oxford University Press, 1954), 7.
2. Gaston Bachelard, *The Poetics of Space*, 2nd edn. (Boston: Beacon Press, 1994), 46.
3. Neutra, *Survival through Design*, 18 (emphasis added).
4. Ibid., 17.
5. Patricia Churchland, *Neurophilosophy: Toward a Unified Philosophy of Mind and Brain* (Cambridge, MA: MIT Press, 1986).
6. Emily Badger, "Corridors of the Mind," *Pacific Standard*, November 5, 2012.
7. Neutra, *Survival through Design*, 7.

1

"KNOW THYSELF": OR WHAT DESIGNERS CAN LEARN FROM THE CONTEMPORARY BIOLOGICAL SCIENCES

Harry Francis Mallgrave

It is no secret that architects generally pride themselves on being artists. Design education begins with elementary artistic exercises, and professional and nonprofessional journals today consider the architect to be an important arbiter of taste within the arts, in the same way that a generation or two ago the work of Mondrian or Matisse personified the idea of a transformative modernity. But wherein resides the "art" within the "art of building" (*Baukunst*)? I do not raise this question disparagingly, but I do mean to be provocative. Does the artistic component of an architectural design reside in its creativity, composition, good forms, functionality, daring structure, intentionality, or the grand ensemble of a self-satisfying object plopped within a particular context? Or does it lie elsewhere? Architects generally avoid this question.

The sociologist Tim Ingold has noted something else curious about some architects. He points out that many designers and other plastic artists today often define themselves precisely through the fact that they have transcended the "technical" side of their profession. Their chief challenge, it seems, is to exercise their imaginations, whereas the presumably less imaginative technicians or minions in the field apply their tools to construct the artifact "dreamed up" by the architect.[1] This fact seems odd in another respect, because the word architect, of course, comes from the Greek word *architekton*, which means "chief craftsman" or "master builder." And the Greeks similarly divided the arts into two categories. At the top were the celebratory arts of music, drama, poetry, song, and dance, which were lovingly bestowed upon humanity by no lesser deities than Dionysus, Apollo, and the Muses. Below this group was a second tier consisting of the "technical" arts of painting, sculpture, and architecture, which had little in the way of divine sanction. In fact they owed their more modest and artisanal origin to Prometheus, who—as the Platonic interlocutor Protagoras noted—stole them from the workshops of Hephaestus and Athena. Again, these lesser (although not necessarily less impressive) arts were defined by the cultivation of skills, rather than by the exercise of imagination.

1.1 Gypsum Assyrian panel from the Palace of Ashurnasirpal II, Nimrud, 9th century BCE. British Museum, London.

Discerning the location of the "art" of architecture has another problem, which is the fact that until the eighteenth century at least, architecture was considered as much a science as an art. And throughout recorded history architects turned to the sciences to find their philosophical or aesthetic footing. This was certainly true of the classical tradition in its various guises. In book one Vitruvius famously stressed that the architect (in addition to having a knowledge of drawing, philosophy, and history) should be well versed in geometry, optics, arithmetic, acoustics, musical harmony, medicine, law, and astronomy, in addition to having expertise in mechanics and hydraulics, and building catapults and siege machines. For the Renaissance theorist Alberti the architect was no less *uomo universale* functioning within a larger cosmological system in which beauty, that elusive mistress of the Vitruvian triumvirate, was accessible only by initiation into the tenets and methods of the sciences. Francis Bacon, at the start of the seventeenth century, placed architecture alongside perspective, music, astronomy, cosmography, and engineering as one of the "mixed sciences," and many of the major architects of that century, such as François Blondel, Claude Perrault, Christopher Wren, and Guarino Guarini, could be equally viewed as scientists.[2] The anatomist Perrault, who was a member of the French Academy of Sciences and not the French Academy of Architecture, famously untethered architecture from its Albertian macrostructure not by aesthetic decree but by arguing that the eye and ear process their respective sensory stimuli in very different ways, physiologically.[3] The line dividing science and architecture at this time was virtually indistinguishable.

1.2 Library of Celsus, Ephesus, Turkey, completed 135 CE.

Yet even with architecture's gradual emergence during the eighteenth century as an autonomous artistic discipline, it remained drawn to science for both guidance and aesthetic inspiration. In 1757 Edmund Burke plotted a new course for aesthetic theory, by contending that the emotions aroused by beauty and sublimity had nothing to do with numerical proportions or harmonic ratios, but rather with the relaxation and tensioning of the optic nerve.[4] Such a hypothesis allowed Uvedale Price to construct an entire theory of the picturesque, and the architect Julien-David Leroy to attribute the charm of a moving spectator's experience with a colonnade to the successive display of light and shadow parading across the retinal field.[5] In joining such an idea with traditional French theories of character, Le Camus de Mézières went so far as to suggest that all architectural forms could be read emotionally from the modulation of a building's lines.[6]

Architectural theory continued to draw much of its inspiration from the biological sciences in the nineteenth century. The philosopher Arthur Schopenhauer, in following the lead of Immanuel Kant, argued that perception is no passive process, but one in which the brain actively constructs its world through a complex series of neurological operations. Architecturally, he translated this postulate into the brain reading a building's forms as a conflict between gravity and rigidity. The architect's task was to devise an ingenious system of columns, beams, joists, arches, vaults, and domes, through which he deprives these "insatiable forces [of gravity] of the shortest path to their satisfaction."[7]

Such an interpretation inspired a whole body of architectural theorizing. Karl Friedrich Schinkel, for instance, admitted his earlier "error or pure radical abstraction" with regard to his utilitarian treatment of building forms, while Karl Bötticher saw architecture expressly as a symbolic process, one in which its ornamental vocabulary must be crafted into a language to represent these gravitational forces as an "ideal organism."

1.3 Egyptian basket column capital, from Gottfried Semper, *Der Stil in den technischen und tektonischen Künsten.*

Gottfried Semper's large body of theory, assembled for the most part in the 1850s, was built upon similar premises. If, like Bötticher, he came to see the convex curvature of the Doric echinus as expressing the weight of the load bearing down on the column, he interpreted the "supple and elastic strength" of an Ionic volute as offering "resistance without violence."[8] The concave curvature of an Egyptian basket-weave capital, by contrast, came about because the strong textile fibers were in a state of tension, restraining the outward force of the load bearing down on the capital.

Semper's theory also marks a time in which architectural theory aligns in an interesting way with physiological research. His friend and colleague at the ETH in Zurich, Friedrich Theodor Vischer, similarly viewed the architect's task as one of infusing "buoyant life" into inert matter through a linear and planar suspension of forms. Pointing to contemporary medical research, he argued that this animistic effect was due to the fact that all forms induce "certain vibrations and who knows what neural modifications" within the spectator's organism.[9] In 1873, his son Robert coined a term to define this process of neural modification—the German word *Einfühlung*. It is a difficult word to translate, and while the English word "empathy" is adequate for many purposes, for the younger Vischer it connoted the active process by which we literally "feel" ourselves into or neurologically simulate objects of artistic contemplation. When we experience a great work of art, we not only have an "intensification of sensuousness" but also a general strengthening of our vital sensations. "Every work of art," he noted, "reveals itself to us as a person harmoniously feeling himself into a kindred object, or as humanity objectifying itself in harmonious forms."[10]

Vischer's relatively small book touched off a veritable deluge of writings on aesthetic "empathy" in the last decades of the nineteenth century, culminating along one front with Heinrich Wölfflin's remarkable dissertation of 1886, which opened with the question "How is it possible that architectural forms are able to invoke an emotion or a mood?"[11] I will return to aspects of this work later in this chapter, but I want to point out one other bold attempt to realize this line of aesthetic reasoning—the founding of the new garden city of Hellerau in 1906.

One of its co-founders was Wolf Dohrn, the son of a prominent biologist who in 1902 had completed his dissertation under the famed psychologist of empathy theory, Theodor Lipps. Dohrn proposed various ways to make this new "German Olympus" an Edenic center of social good will and personal enhancement, but none was more revolutionary than to make musical training the centerpiece of its educational system. To this end he persuaded the famed musicologist Émile Jaques-Dalcroze to relocate his conservatory to Hellerau. The aim of the institute, however, was not musical competence in itself but to employ it as a means to enhance a person's happiness and creativity. Underlying this training was the hypothesis that the movements of the body should be in alignment with the neural activities of the brain, to bring about a "co-ordination between the mind which conceives, the brain which orders, the nerve which transmits and the muscle which executes."[12] In short, it was a holistic theory of embodiment, one that—to frame it within the context of today's biological viewpoint—views consciousness, thinking, and communication activities as intrinsic to our biological organisms or functioning bodies, responding of course to the characteristics of the physical, social, and cultural environments in which we dwell.

Dalcroze's system, integrated with other activities at Hellerau, created a sensation across Europe, and its music and dance festival of 1913 alone attracted 5000 visitors from across the globe. Among the intellectuals who came to Hellerau to evaluate its program in its first few years were Ebenezer Howard, Martin Buber, George Bernard Shaw, Max Reinhardt, Serge Diaghilev, Thomas Mann, Stefan Zweig, Oskar Kokoschka, Emil Nolde, Hugo Ball, Heinrich Wölfflin, Max Klinger, Wilhelm Worringer, Julius Meier-Graefe, Franz Kafka, and Upton Sinclair. Architects were no less interested in the experiment. Peter Behrens, Henry van de Velde, Hans Poelzig, and others involved with the German Werkbund closely followed the events of the town. Walter Gropius (whose future wife Alma Mahler was a frequent visitor) was certainly aware of the theoretical underpinnings of the program, and Mies van der Rohe surely visited his fiancée Ada Bruhn, who studied at Dalcroze's institute in 1912–1913. Charles-Édouard Jeanneret (later Le Corbusier) visited the town on four occasions, in part because his brother Albert was an instructor under Dalcroze. Had this experiment not been rudely halted by the

1.4 Émile Jaques-Dalcroze.

confluence of events of 1914—the outbreak of World War I, Dalcroze's return to Switzerland, and Dohrn's death in a skiing accident—the course of European modernism might well have been very different.

Nevertheless, there was an afterglow to these events in the 1920s. The Soviet Constructivist and De Stijl movements all bore the imprint of many of the earlier cognitive and perceptual experiments, and Hellerau's theories of empathy were particularly evident in the teachings of the Bauhaus, in the embodied work of Johannes Itten, Gertrud Grunow, Oskar Schlemmer, Wassily Kandinsky, and László Moholy-Nagy—all too little discussed within the extensive literature of this period. These efforts, however, were decidedly on the wane with the transfer of the Bauhaus from Weimar to Dessau, and by the early 1930s avant-gardism had all but collapsed across a European continent about to plunge into the cauldron of yet another war. Efforts to merge artistic production with biological knowledge became more infrequent. One exception was Richard Neutra, and his book *Survival through Design*, published in 1954, was largely written during the war years. In this remarkable study, he passionately urged architects to incorporate "current organic research" and "brain physiology" into their designs, to explore our multisensory interaction with the built world, as well as to undertake research in areas of "sensory significance."[13]

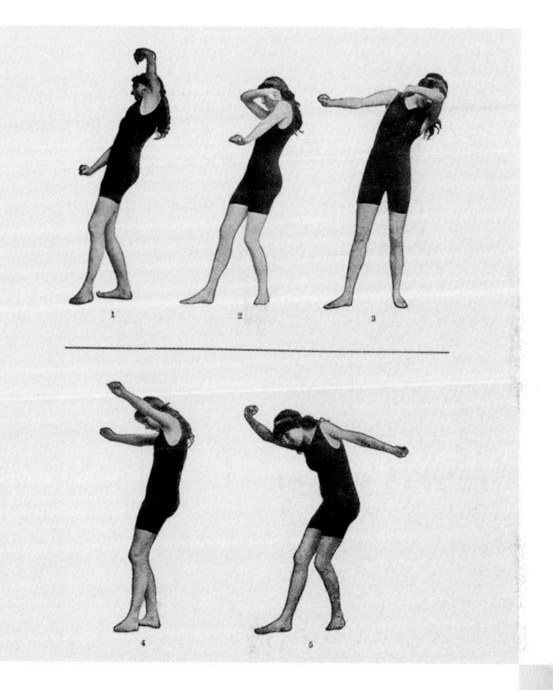

1.5 Eurhythmic exercise at Hellerau, "Beating 5/4 in canon with expression."

There is little need to dwell on the more familiar patterns of architectural theory in the second half of the twentieth century. On the one hand there was a meliorist postwar belief that architects could affect social change, and that this amelioration would take place specifically through technology and its applications. On the other hand there were the more cynical winds of a gathering poststructural storm that—in the effort to preserve architecture's presumed autonomy and theoretical purity—branded all such efforts as "metanarratives." In the ensuing conflict between these competing camps, the phenomenological efforts of Steen Eiler Rasmussen and Christian Norberg-Schulz, the urbanism of Kevin Lynch, the sociological patterns of Christopher Alexander, and the Gestalt thinking of Rudolf Arnheim gained little traction. The political abstractions of the

Frankfurt School, semiotics, Lyotard, Foucault, and "weak thought" all proved to be far more interesting, and when the last Derridean converted to Deleuzianism in the early 1990s, architectural theory had all but crashed and burned in its conceptual excesses. By this time architects had quite rightly grown weary of "isms," and in any case the profession was passing into the digital and green age, when the spectacle of all such strategies was presumed to be irrelevant. Sustainability, new and improved building materials, parametric softwares, and building information modeling (BIM), it was presumed, would constitute the future of design.

ΓΝΩΘΙ ΣΕΑΥΤΟΝ (KNOW THYSELF)

These two words, according to Pausanias, were inscribed in stone at the entrance of the Temple of Apollo at Delphi.[14] The maxim, he reports, was written by one of the wise men of ancient Greece, and Plato in particular was fascinated with the phrase. He refers to it in no fewer than six of his dialogues, and in two instances Socrates invokes the phrase to caution those who in their vanity seek to understand obscure and far-flung knowledge without first understanding their own human natures.[15] I invoke the admonition in this sense, because I believe that what has been lacking in so much of architectural thought of the past half-century has been a concern for ourselves—not only how people actually experience the built environment, but also how the "inner voice" of the designer might align itself with the social institutions and cultural edifices of our explosive metropolises. Somewhere within this reconciliation, I will argue, lies the "art" of design.

I want to focus on the biological sciences, but I want to begin with a small digression into ethology, the science dealing with animal behavior and its evolutionary transformations. For some years Ellen Dissanayake has been seeking the origin of the arts, which she has done by drawing upon evolutionary and anthropological evidence, by exploring the human predisposition for play, as well as for ritualistic and ceremonial behavior. In her book *Art and Intimacy*, which came out in 2000, Dissanayake buffers her case by citing the work of Colwyn Trevarthen, a child psychologist and psychobiologist at the University of Edinburgh. Some of his experimental studies focused on how mothers and infants build loving bonds through such measures as "baby talk"—that is, the "cooing" patterns of intonation, vocal and rhythmic exaggerations, and repeated visual and tactile give-and-takes. In her book, Dissanayake draws a far-reaching conclusion from this research:

> I show that human newborns come into the world with sensitivities and capacities that predispose them to join in emotional communion with others. I then argue that these same sensitivities and capacities, which arose as instruments of survival in our remote hominin past, are later used and elaborated in the rhythms and modes of adult love and art.[16]

Such a thesis is intriguing, especially since neuroimaging studies have since demonstrated that beauty (in art) and love (romantic and maternal) do indeed share a similar hedonic or "pleasure circuit."[17] Dissanayake goes on to expand her argument along several fronts. First, these "rhythms and modes" underlying artistic expression, which extend back into pre-Paleolithic stages of human evolution and are enacted from our first minutes of life, are presymbolic in their biological underpinnings. The human artistic impulse therefore seems to be more deeply rooted in our human natures than the speculative anthropological theories of the past have allowed. Second, they are related to emotional drives associated with enculturation, such as social affiliation, making sense of our surroundings, acquiring competence in skills, and what she refers to as "elaborating upon." In this last regard, the arts "emerged through human evolution as multi-media elaborations of rhythmic-modal capacities that by means of these elaborations gave emotional meaning and purpose to biologically vital activities."[18] And it is only when the artist or architect taps into these "cross modal sensations of tactility and kinesis" together with their emotional use of colors, forms, and textures, that a work of art or architecture attains the charm of being both creative and revelatory, that is, of becoming something "special."

Dissanayake's hypothesis has profound implications in many respects, which space does not allow us to pursue here. I raise her hypothesis for the simple reason that today the biological sciences are indeed telling us much about ourselves, much more than we have learned in the past. And while architectural theory over the past half-century has been pursuing the phantom of its coveted autonomy, the biological sciences and their sister disciplines in the humanities have been spawning a bevy of interdisciplinary fields that have been decoding the mysteries of human life with unparalleled success. In 1949 Donald Hebb concluded that when two neurons fire together, protein growth takes place and the synaptic bond between them is strengthened.[19] This hypothesis, confirmed only some years later, led to our present understanding of learning and neural plasticity. In 1953 James Watson and Francis Crick demonstrated that the double helix was the configuration of the DNA structure, which laid the basis for decoding the human genome. In the 1970s and 1980s, biological pioneers such as Benjamin Libet and Gerald Edelman began to probe the phenomenon of how human consciousness arises. In the early 1990s a team of researchers in Parma, led by Giacomo Rizzolatti, discovered "mirror neurons" in the brains of macaques.[20] In the same decade the Human Genome Project was initiated, which resulted in 2003 in the full sequencing of the 25,000 or so genes of the human genome. During these same two decades we saw the proliferation and continuing refinement of neuroimaging technologies, such as positron emission tomography (PET scans) and function magnetic resonance imaging (fMRIs), which have allowed us to record neurological aspects of the working human brain. Today we have even isolated particular areas of the brain involved with the perception of buildings, landscapes, and their spatial properties. It is no exaggeration to say that we have learned more about our biological selves in the past half-century than in all of human history, and as a result of these developments, the humanities—sociology, philosophy, psychology, and human paleontology in particular—have been forced to restructure their premises and research agendas radically. Yet architects have remained surprisingly incurious or seem little moved by these events.

I believe this incuriosity is no longer tenable. For one of the new perspectives that is emerging from these breakthroughs is the simple recognition that we are developmental organisms raised within environmental fields, and that the quality of these environmental fields has a powerful impact on our cognitive and organic development over a relatively short time. And if we accept the premise that architects are the principal designers of our built environments, it would seem to be incumbent upon them to learn something about our biological complexity. Here I am not referring to the obvious hazards to human health and welfare that have been well publicized, such as the dangers of pesticides in our food or the presence of formaldehyde in our habitats; rather, I am speaking of the larger and more complex manner in which we take in or experience the built world, how it indeed might give us some pleasure or happiness, or rest and refuge, as our needs may dictate. The investigations undertaken thus far by the biological sciences are already rich in their implications, but for architecture, at least, they remain to be fully explored. Nevertheless, let us focus on two aspects that have direct relevance to our built environment. The first derives from the new models of emotion that are presently evolving, and the second is emotion's underpinning in the mirror neuron system. Both aspects become more acute as we turn over more and more of our design responsibilities to machines.

EMOTION

Any student of architectural ideas, particularly someone bred on the philosophical abstractions of recent theory, might be surprised if not bemused to stumble upon this term. For the word "emotion" has rarely been raised within polite architectural circles for the past century or so, and it is certainly fair to say that the idea has never formed a significant part of theoretical discourse since the picturesque or *architecture parlante* movements of the eighteenth century. It is as if the very notion is an anachronism that we, with our abstract powers of logical reasoning, have pretended to have outgrown or moved beyond. But why indeed are architects so distrustful of this word with respect to design?

The famed biologist Joseph LeDoux defines emotion as "the process by which the brain determines or computes the value of a stimulus."[21] Emotions can be of many varieties, and I use the term here in a nonspecific sense—that is, not as someone's heartfelt response to a particular building but rather as the way that we more broadly connect with or experience our built environments. These connections take place in a variety of ways, but the key fact is that emotion, as emphasized in the perceptual theories of James Gibson, is always a multimodal or multisensory experience of someone moving through an environmental field.[22] We engage the world holistically on all sensory levels, and the sum of these sensory contacts initiates what Jaak Panksepp refers to as "affect" programs—that is, electrochemical responses generated in the subcortical areas of the basal ganglia and brainstem (among other places), triggering chemical responses in other areas of the brain.[23] Douglas Watt notes that "emotion probably reflects a special kind of carefully routinized and evolutionarily carved set of operations for protecting our homeostasis."[24] Broadly speaking, emotion conditions our response to specific events or sensory fields. In this sense, emotion is precognitive or prereflective because its triggering, by most definitions, takes place prior to our conscious awareness of this subcortical activity. Emotions are embodied within our perceptions, and it is only later that we reflect upon our "feelings" toward some event.

Emotion, as we noted, can be of many types and complexity. Panksepp, for instance, refers to "seeking" and "play" as two emotional endophenotypes, ingrained behaviors induced by the interaction of genes and the environment.[25] A young bird must at some point leave its nest and explore the world in order to survive, and the human biological system similarly needs stimulation and information to prosper. Thus novelty is one motivational factor that often entails a certain pleasure, especially when it informs us with a new perspective. Similarly, physical play is something that strengthens muscles, enhances the metabolic system, and leads to the cultivation of motor skills; it also allows mammals to bond socially and, in an artistic sense, it is an essential aspect of creative thinking. When seeking and play give us joy, it is because they set in motion that neural "pleasure circuit" that, once ignited, floods our brains with a mixture of neurotransmitters, such as dopamine. Traditionally, philosophers have characterized emotion as something opposed to reason, but this is an antiquated way of viewing things. Emotion is the multisensory medium through which we engage the world, and human reason is but an evolutionary refinement of the emotional process—circuits that delay the animal response pending "further review." Yet emotion has already set the tenor of not only what is to be reviewed but how it is to be reviewed, and in this way it is powerfully implicated in our responses to the built environment.

The newer models of emotion are important to architecture for two reasons. First of all, they suggest that our initial emotional engagement with the environment is precognitive or nonconscious. Much of this neurological activity (tens of billions of neurons) operates below the threshold of conscious awareness for the simple reasons that it is too massive an activity and operates too quickly for the conscious mind to focus on it. Secondly, awareness and thought are fundamentally embodied, in the sense that important aspects of our perceptual and conceptual activity involve the sensorimotor areas related to our movements and corporeal awareness. What this means, in the words of George Lakoff and Mark Johnson, is "that human concepts are not just reflections of an external reality, but they are crucially shaped by our bodies and brains, especially by our sensorimotor system."[26]

Yet architects sometimes go to great lengths to spin elaborate yarns to intrigue or beguile the patron or public, usually with the presumption that they are informed readers of the script. Whereas allegory and metaphor are perfectly valid exercises in architectural design, architects at times need to be reminded that people initially do not experience their habitats in this way. The general ambience of a perceptual field is what people first encounter, in large part through our peripheral vision, as Juhani Pallasmaa has noted. And biological judgments are already being made by such things as the touch of a door handle or handrail, the proportioning of stair risers and treads, the texture of the floor material, the resonance or ambience of the spaces, the hand of fabrics, the smell of materials, and the presence of natural light. These biological responses occur before someone stands back and reflects on the overall experience.

There is nothing radically new in such a statement. The better architects of the past and present have always been aware of this fact. What biology today is making evident, however, is the degree to which our responses to physical, social, and cultural environments are embodied, and how our responses to this condition of embodiment in turn alters our biological organisms. We "feel ourselves into" (to use Robert Vischer's term) our living environments in a multisensory and immediate way through our bodies, and these feelings have biological consequences. Yet the profession and teaching of architecture has for some years been moving in a contrary direction, as the notion of what constitutes good design is reduced more and more to iconic or novel images found on the web, computer-generated renderings of yet-to-be-built designs instantaneously made available to all. This degradation of the sensory aspects of architecture is particularly problematic with students who are quite naturally fascinated with the power to wield or manipulate forms endlessly on a computer screen. Yet design is indeed a zero-sum game. The aspects on which one focuses one's effort during the design process determine, to a great extent, what the final result will be. When one devotes an inordinate amount of one's attention to compositional or novel form, for example, one tends to ignore materiality and detailing. When one seeks out only novelty, one tends to ignore historical examples that might offer other important lessons in design. And even when one is sensitive to the fact that architecture has a history, it is often expressed in superficial ways. The image of a Brunelleschi church in a history book does not prepare the student for the actual experience of leaving the Florentine summer heat and walking down the spatial expanse and thermal coolness of the nave of Santo Spirito.

Emotional research also offers us a refrain from the excesses of recent architectural theory. At the risk of belaboring this point, the poststructural notion that architecture should mirror our "decentered condition" or express our existential angst overlooks one important thing. People primarily evaluate their ecological environments through the ensemble of stimuli generated by the materials selected, spatial relations, formal proportions, scale, patterns, rhythms, tactile values, and creative intentions—not to mention those more arcane matters of comfort, convenience, craftsmanship, presence, warmth, and beauty. The word "aesthetics" indeed comes from the Greek word *aisthētikos*, which simply means "sense perception" or "to perceive."

On an urban level, one of the insights of these newer emotional models is the recognition that our emotional responses are strongly integrated with our peripheral autonomic nervous system—that is, the working of our sympathetic and parasympathetic subsystems. These neural subsystems work in a reciprocal and opposing fashion. The sympathetic system, for instance, may accelerate the heart rate in response to one bodily condition, while the parasympathetic subsystem may slow it down in response to another condition. These two subsystems, in turn, are separately wired into the insular cortices in each hemisphere of the brain (a cortical region behind each ear yet tucked toward the center of the brain). The sympathetic subsystem terminates in the right insula and, as A. D. "Bud" Craig has reported, is associated with energy expenditure and arousal. Conversely, the parasympathetic subsystem terminates in the left insula and is a response to energy nourishment, relaxation, and affiliative emotions.[27] The insula is now recognized as one area of the brain in which we become aware of our visceral and emotional feelings.

This fact is important because the built environment can be aligned with these two poles. A building can arouse our metabolic systems and demand high energy expenditure, or a building can provide a place of relaxation and comforting sociability. Some buildings or environmental events can do both, but my point is the very obvious fact that we can approach a design problem in two general ways. We can design for the "wow" effect, the highly stimulating environment that forces people to come to terms with the intensity and presumed ingenuity of our design, or with greater modesty we can design a place that provides rest and comfort, or perhaps offers the occasion for social rituals or private nourishment. Both approaches have their appropriate occasions, but then again architectural training in the design studio often encourages the seeking out of the greatest "bang for the buck," as the saying goes. With this in mind, what happens in an urban environment where most buildings are designed to be active, aggressive, or even abrasive to our senses? Major squares and avenues in New York or Tokyo may be rightly celebrated for this attraction, but it also stands to reason that cities—as biophilic proponents have long argued—should also have greener or more nature-centered areas to maintain a sensory and psychological balance. Biological experiments related to architecture will in the future no doubt refine our approach to design in this regard. For instance, areas in schools can be active or passive depending on the educational or recreational tasks at hand, and indeed three-dimensional immersive technologies are already allowing us to study people's responses to such environments before they are constructed.

MIRROR SYSTEMS

Emotions function with such unerring immediacy in part because of another major biological discovery of the 1990s—that of mirror neurons or (for humans) mirror systems.[28] The discovery may very well prove to be one of the most important scientific events of the past half-century, but it should also be emphasized that research on these systems in humans is still in its infancy, and an understanding of their full implications remains years away. In the discovery of the early 1990s, scientists inserted electrodes into the brains of macaques to record the neural circuits involved with grasping objects such as peanuts. What they found to be unusual was that certain neurons became active in monkeys who were not grasping things, but were simply watching others grasp objects. Neuroimaging technologies have demonstrated similar "mirror systems" in humans. What is also interesting is that one of the areas in humans presumed to contain mirror neurons for grasping is the premotor cortex, which is involved with the production of speech. This fact in itself has upended a few thousand years of linguistic (not to mention philosophical) theory, for it suggests that human speech was built upon more ancient brain mechanisms that were involved with action recognition and manual gesturing.[29]

The discovery of mirror neurons and mirror systems has led to hundreds of neuroimaging studies over the last fifteen years surveying their extent and significance. We now know, for instance, that there is not only a mirror system for grasping in humans, but several such systems distributed across different brain regions. Mirror systems are also active with emotional responses, and thus are now presumed to be part of the reason we have social empathy.[30] It has also become clear that in witnessing someone in pain, we map the area of trauma onto our own bodies.[31] Again, mirror systems seem to be active in proprioception, in the sense that we seem to enjoy the movements of a ballet dancer not just visually but also motorically. It is as if the motor circuits in our own brains, in mirroring the motor movements of others, take a simulative pleasure in ourselves moving with such agility and poise.[32] From a different perspective, some human brain disorders, such as autism, are now presumed to result from a breakdown of a mirror system. Much of our social understanding of mirror neurons has been collected under that old term "empathy" theory, but another term gaining traction today for mirror systems is "embodied simulation."

Mirror systems have more recently been implicated in artistic events, and in this regard two studies are of importance to architects. In the first one, scientists were recording the activity of the mirror system involved with touch, such as watching someone touching another person. We might expect such a response because of our social empathy, but the scientists also found evidence of mirror activity when we observe two inanimate objects touching one another. As the neuroscientists framed the issue: the domain of touch appears not to be limited to the social world. Space around us is full of objects accidentally touching each other, that is, without any animate involvement. One could observe a pine cone falling on the garden bench in the park, or drips splashing on the leaves of a plant during a downpour. Models of embodied simulation posit that the same neural structures involved in our own body-related experiences contribute to the conceptualization of what we observe in the world around us.[33]

Granted that we may not generate an empathic accord with every aspect of our inanimate environment, this simulation of features of the inanimate world—which of course encompasses the built environment—returns us to the empathy theories of Semper, Theodor and Robert Vischer, and Wölfflin. One of Wölfflin's criticisms of Robert Vischer's proposal was that he was envisioning empathy as a kind of psychological projection of our self into the artistic entity, through which we read back our emotional responses or feel our own experience. Wölfflin insisted that we animate architectural events simply "*because we ourselves possess a body*"—that is, because the optic nerve stimulates the motor nerves and thereby sympathetically works on our own neural system through our bodily organization.[34] Because we know the force of gravity through our own corporeal

1.6 Leon Battista Alberti, Santa Maria Novella, Florence, 1456–1470.

experience, we read the weight and balance of a building in gravitational terms. We judge a work of architecture to be beautiful because it in fact mirrors the "*basic conditions of organic life.*"[35]

Wölfflin was not alone in such theorizing in his day. Around the turn of the century the English writer Vernon Lee, in collaboration with the painter Clementina Anstruther-Thomson, attempted to record, through a series of experiments, the physiological responses of people viewing buildings. In studying the façade of the church of Santa Maria Novella, for instance, the two artists recorded how its proportions altered and moderated breathing patterns, exerted certain pressures on the feet and head, and ultimately uplifted the viewer with a feeling of "harmonious completeness."[36]

All of this leads us to the second recent study that very much relates to the experience of architecture. In 2007 the art historian David Freedberg, in collaboration with the neuroscientist Vittorio Gallese (one of the discoverers of mirror neurons in Parma), argued that the experience of art and architecture worked through the precognitive activation of embodied mirror mechanisms involved with the simulation of actions, emotions, and corporeal sensations. This contention, one might argue, is not unexpected with figurative

1.7 *Laocoön*, Late Hellenic work attributed to the Rhodian sculptors Agesander, Athenodoros, and Polydorus, ca. 42–20 BCE. Vatican Museums.

works of art in which we form some powerful emotional or empathic attachment, such as the well-known Hellenic sculpture of Laocoön and his sons being murdered by the wrath of Athena. But the authors go one step further. They argue that we can also read "the visible traces of the artist's creative gestures, such as vigorous modeling in clay or paint, fast brushwork and signs of the movement of the hand more generally."[37]

One of the examples that the authors present is how viewing any of Michelangelo's *Prisoners* often prompts the response of activating muscles within our own systems, as if we, like the prisoners, were struggling to free our body from the stone. Another example is how a twisted column might induce a state of tension within our bodies, as our mirror systems viscerally simulate the twisting of the column. In the case of the twisted columns and piers in the church of the Monastery of Jesus, circa 1498, such simulation can be read both symbolically and emotionally. Symbolically, the twisting visually strengthens the supports for assuming the load of the heavy vaults, while emotionally this tense gesture seems entirely appropriate in a chapel that was designed specifically to house the ritual sacrifice of Christ.

1.8 Diégo Boytac, Church of the Monastery of Jesus, Setúbal, Portugal, c. 1498.

Or let us take the alabaster bas-relief of the Assyrian warrior shown in the frontispiece to this chapter. In conventional art history books, we read these panels as a narrative depicting the proud warrior in victory, someone who has just dispatched his enemies and brought greater glory to his nation. We may indeed read it in this manner, as many people do, but when we go to the British Museum to study this panel we also find ourselves reading it in a different light. We study the delicate chisel marks that create the composition; we admire the intricacy and detail of the author's hand, the skill that is always present in a great work of art. We are simulating our own hand carrying out this work.

In this way we can now understand how Semper and Wölfflin read the fluted Doric column as a powerful assertion of upward force, a force that is only temporarily contained by the architrave before it breaks through and is re-represented in the triglyphs, and then goes on to lift the gable line of the pediment—at least in Wölfflin's view.[38] Wölfflin also once defined ornament in a very insightful way as "*an expression of excessive force of form.*"[39] Embodied simulation, once again, helps us to understand why

1.9 Parthenon, Athens, east façade, 447–432 BCE.

Semper felt that the chisel marks of a rusticated wall should always be collected toward the center, because in this way the forces of the hammer are contained by the ashlar's band, and the overall rhythmic quality of the wall's "beat" can be preserved. We do in fact read architecture in these ways. If we stand beside the Palazzo Medici in Florence, for instance, we know the weight of these ashlars and we secretly admire the skill it took to hoist these blocks into place. A Japanese Zen shrine, by contrast, overwhelms us with its feeling of lightness.

Rasmussen once commented on an art historian's description of a town that he had gleaned through photographs. Rasmussen knew that we do not experience a city in the same way that we experience images. When we physically come to a town we observe or feel its overall character, atmosphere, topography, sounds, colors, scale, odors or fragrances, and material presences.[40] We in fact judge every building with which we come in contact in this way, as bodies moving through material spaces. Our sensory modalities, as neuroimaging studies amply document, are fully connected and integrated with each other. When we view an architectural material, for instance, we know that the tactile areas of the somatosensory cortex are also called into play. In other words, in an act of simulation we at the same time simulate the touch of the surface with our hands, we inhale its odor, we pick up traces of its acoustic resonance or hardness, and, if we are children still in a state of exploration, we might even lick it. With this in mind, we might rightly ask what happens to a

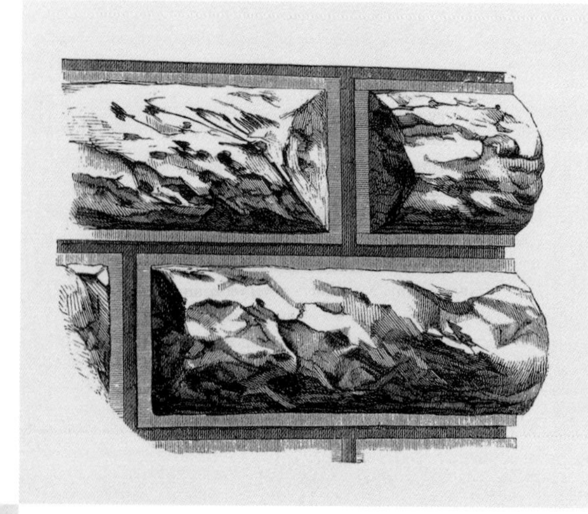

1.10 Rustication detail, from Gottfried Semper, *Der Stil in den technischen und tektonischen Künsten.*

building when it is reduced to a disembodied abstraction formed by algorithms? How does a student today learn a material's tactile values or the smell of cedar?

This is not the place to address these questions, but then again the new interdisciplinary fields that have come into view already make it evident that the increasing computerization of the design process, which has come about by paring away almost all class work related to our biological, social, and encultured natures, has a significant price for the profession at large. Do we really want yet another team of consultants on every large project to inform architects of their fundamental obligations as designers? Would it not be better to integrate these new insights into our curricula? For too many years now we have viewed architecture as a speculative exercise clothed in philosophical abstractions, as the creation of self-satisfying objects under the guise of aesthetic hypotheses that are, quite frankly, obsolete. It is time to reevaluate the matter, and perhaps employ the new tools at our disposal to explore other aspects of our beings, such as those "rhythms and modes" of which Dissanayake spoke. What are these empathic systems underlying our sociability, and how can they be clothed in suitable architectural forms? Why is it that we seem to have this innate capacity for acquiring skills or appreciating craftsmanship, and again, what are the architectural implications for such drives? The substance of the new research will not in any way inhibit technological advances or undo creative thinking. In fact I believe the opposite will be the case, because these new models will offer architects a means to rethink their tasks and provide design with a more secure theoretical footing. Knowing ourselves, as Socrates would undoubtedly agree, will help us better to know the people for whom we build. Perhaps it will provide us with the means for rediscovering the location of architecture's missing "art."

NOTES

1. Tim Ingold, *The Perception of the Environment: Essays on Livelihood, Dwelling and Skill* (London: Routledge, 2000), 349–351.

2. Francis Bacon, *The Works of Francis Bacon*, Book II (London: J. Johnson et al., 1803), 108.

3. Claude Perrault, *Ordonnance for the Five Kinds of Columns after the Method of the Ancients* (Santa Monica: Getty Publication Programs, 1993), 49.

4. Edmund Burke, *A Philosophical Enquiry into the Origin of Our Ideas of the Sublime and Beautiful* (1757).

5. See Uvedale Price, *Essays on the Picturesque as Compared with the Sublime and the Beautiful* (London: Mawman, 1794); Julien-David Le Roy, *Histoire de la disposition et des formes différentes que les chrétiens ont données à leurs temples* (Paris: Desaind & Saillant, 1764), 63.

6. See Nicolas Le Camus de Mézières, *The Genius of Architecture; or, the Analogy of that Art with Our Sensations*, trans. David Britt (Santa Monica: Getty Publication Programs, 1992).

7. Arthur Schopenhauer, *The World as Will and Representation*, trans. E. F. J. Payne (New York: Dover, 1969), §43, 214–215.

8. Gottfried Semper, *Style in the Technical and Tectonic Arts*, trans. Harry F. Mallgrave and Michael Robinson (Los Angeles: Getty Publications, 2004), 783.

9. Friedrich Theodor Vischer, "Kritik meiner Ästhetik," in *Kritische Gänge* (Stuttgart: Cotta, 1866), 5: 143.

10. Robert Vischer, "On the Optical Sense of Form," in *Empathy, Form, and Space: Problems in German Aesthetics, 1873–1893*, trans. Harry F. Mallgrave and Eleftherios Ikonomou (Santa Monica: Getty Publications, 1994), 117.

11. Heinrich Wölfflin, "Prolegomena toward a Psychology of Architecture," in *Empathy, Form, and Space* (note 10), 149.

12. Émile Jaques-Dalcroze, "Rhythm as a Factor in Education," in M. E. Sadler, ed., *The Eurhythmics of Jaques-Dalcroze* (Boston: Small Maynard, 1939), 18.

13. Richard Neutra, *Survival through Design* (New York: Oxford University Press, 1954), 111–244.

14. Pausanias, *Guide to Greece*, Book 10:24.

15. Plato, *Phaedrus*, 229, and *Philebus*, 48.

16. Ellen Dissanayake, *Art and Intimacy: How the Arts Began* (Seattle: University of Washington Press, 2000), 6. For her discussion of the work of Colwyn Trevarthen, see 15–17.

17. See Semir Zeki and Andreas Bartels, "The Neural Basis of Romantic Love," *Neuroreport* 11, no. 17 (2000): 3829–3834; Hideaki Kawabata and Semir Zeki, "Neural Correlates of Beauty," *Journal of Neurophysiology* 91 (2004): 1699–1705; Semir Zeki and Andreas Bartels, "The Neural Correlates of Maternal and Romantic Love," *Neuroimage* 21 (2004): 1155–1166.

18. Dissanayake, *Art and Intimacy*, 145. See also my discussion of her theories in *Architecture and Embodiment: The Implication of the New Sciences and Humanities for Design* (Abingdon, UK: Routledge, 2013), 192–203.

19. Donald Hebb, *The Organization of Behavior: A Neuropsychological Theory* (New York: John Wiley, 1949).

20. See Giacomo Rizzolatti and Corrado Sinigaglia, *Mirrors in the Brain: How Our Minds Share Actions and Emotions*, trans. Frances Anderson (New York: Oxford University Press, 2008).

21. Joseph LeDoux, *Synaptic Self: How Our Brains Become Who We Are* (New York: John Wiley, 2000), 206.

22. James Gibson, *The Ecological Approach to Visual Perception* (Hillsdale: Lawrence Erlbaum Associates, 1986).

23. For a succinct summary of his ideas, see Jaak Panksepp, "On the Embodied Neural Nature of Core Emotional Affects," *Journal of Consciousness Studies* 12, no. 8 (2005): 158–184.

24. Douglas F. Watt, "Panksepp's Common Sense View of Affective Neuroscience Is Not the Commonsense View in Large Areas of Neuroscience," *Consciousness and Cognition* 14, no. 1 (2005): 81–88.

25. Jaak Panksepp, "Emotional Endophenotypes in Evolutionary Psychiatry," *Progress in Neuro-Psychopharmacology and Biological Psychiatry* 30, no. 5 (July 2006): 774–784.

26. George Lakoff and Mark Johnson, *Philosophy in the Flesh: The Embodied Mind and Its Challenge to Western Thought* (New York: Basic Books, 1999), 22.

27. A. D. "Bud" Craig, "Forebrain Emotional Asymmetry: A Neuroanatomical Basis?," *Trends in Cognitive Sciences* 9, no. 12 (December 2005): 565–570.

28. For the authoritative summary of the discovery and its implications, see Rizzolatti and Sinigaglia, *Mirrors in the Brain.*

29. See Michael Arbib, *How the Brain Got Language: The Mirror System Hypothesis* (Oxford: Oxford University Press, 2012). See also his chapter in this volume (chapter 4 below).

30. See especially Christian Keysers, *The Empathic Brain: How the Discovery of Mirror Neurons Changes Our Understanding of Human Nature* (Lulu.com, 2011).

31. See Tania Singer et al., "Empathy for Pain Involves the Affective but Not Sensory Components of Pain," *Science* 303, no. 5661 (2004): 1157–1162.

32. Beatriz Calvo-Merino et al., "Towards a Sensorimotor Aesthetic of Performing Art," *Consciousness and Cognition* 17 (2008): 911–922.

33. Sjoerd J. H. Ebisch et al., "The Sense of Touch: Embodied Simulation in a Visuotactile Mirroring Mechanism for Observed Animate or Inanimate Touch," *Neuron* 42 (22 April 2004): 335–346.

34. Wölfflin, "Prolegomena toward a Psychology of Architecture," 151.

35. Ibid., 160; emphasis added.

36. Vernon Lee and Clementina Anstruther-Thomson, "Beauty and Ugliness," in *Beauty and Ugliness and Other Studies in Psychological Aesthetics* (London: John Lane, The Bodley Head, 1912), 187–189.

37. David Freedberg and Vittorio Gallese, "Motion, Emotion and Empathy in Esthetic Experience," *Trends in Cognitive Sciences* 11, no. 5 (2007): 199. See also M. Alessandra Umiltà, Cristina Berchio, Mariateresa Sestito, David Freedberg, and Vittorio Gallese, "Abstract Art and Cortical Motor Activation: An EEG Study," *Frontiers in Human Neuroscience* 6 (November 2012): 1–9.

38. Wölfflin, "Prologomena toward a Psychology of Architecture," 180.

39. Ibid., 179; original emphasis.

40. Steen Eiler Rasmussen, *Experiencing Architecture* (Cambridge, MA: MIT Press, 1964), 39–44.

2

THE EMBODIED MEANING OF ARCHITECTURE

Mark L. Johnson

Human beings are creatures of the flesh who arrange spaces and physical structures fitted to their bodies. We live in and through our ongoing interactions with environments that are both physical and cultural. The structures we make are loosely adapted to the functions we perform. Some of these functions are necessary for our survival and flourishing, such as working, eating, having shelter, playing, and sleeping. However, we also order our environments to enhance meaning in our lives and to open up possibilities for deepened and enriched experience. In other words, although we are animals evolved for fitness, we are just as much animals with a deep desire for meaning as part of our attempts to grow and flourish.[1] Architecture is ideally located at the intersection of these two complementary aspects of our lives, insofar as the ways we organize space and buildings address simultaneously our need for physical habitation and our need for meaning. I want to reflect on the nature of human meaning-making through architecture, as it bears on both of these fundamental human needs.

EMBODIED HUMAN MEANING

Toward the beginning of *Art as Experience*, John Dewey argues that the key to appreciating the central role of the aesthetic dimensions for all human experience is the recognition that everything important arises from the ongoing interactions of a living creature with its complex environments:

> Life goes on in an environment; not merely in it but because of it, through interaction with it. No creature lives merely under its skin; its subcutaneous organs are means of connection with what lies beyond its bodily frame, and to which, in order to live, it must adjust itself, by accommodation and defense but also by conquest. At every moment, the living creature is exposed to dangers from its surroundings, and at every moment, it must draw upon something in its surroundings to satisfy its needs. The career and destiny of a living being are bound up with its interchanges with its environment, not externally but in the most intimate way.[2]

We live and become what we are only in and through our engagements with our many environments. All our perceptions, feelings, emotions, thoughts, valuations, and actions are thus consequent on our embodied transactions with our physical surroundings, our interpersonal relations, and our cultural institutions and practices. Our capacity to experience, make, and communicate (share) meaning is not just a result of the makeup of our brains and bodies, but depends equally on the ways our environments are structured.

Before we can examine some of the ways architecture enacts meaning, we need to be clear about what is meant by "meaning." This is a very serious problem, because several decades of work in linguistics and the philosophy of language within what is known as the "analytic" (Anglo-American) philosophical tradition have reinforced the mistaken idea that all meaning is a matter of language, so that it is words and sentences alone that have meaning. It is also usually assumed that sentences are meaningful because they express propositions that map onto states of affairs in the world. In short, meaning is thought to be a matter of how words (and sentences) can be "about" the world, and so it is thought to be propositional and truth-conditional in nature.

The first obvious thing to point out about this mainstream conception of meaning as linguistic is that such a view can have almost no significant application to our experience of architecture, or of any other art, for that matter. For the most part, architectural structures are not linguistic signs that could mean in the way that sentences, phrases, and words are thought to have meaning. I submit that such an impoverished view of meaning gives us virtually no useful way of explaining either the meaning or the power of architecture in our lives. Trying to turn buildings into quasi-sentences and parts of buildings into quasi-phrases or quasi-words in order to show that and how they can be meaningful should make it evident that there is something fundamentally wrong, in the first place, with the view of meaning as entirely conceptual, propositional, and linguistic. In short, although one can probably find some very weak analogies between buildings and sentences, the dominant meaning-as-linguistic view turns out to be a profoundly inadequate way to understand the meaning of architecture, or any other art.[3]

Fortunately, we need not be encumbered by such an inadequate and impoverished theory of meaning. Cognitive science research over the past three decades has given us a dramatically richer view of the nature and sources of meaning. The general framework is one suggested nearly a century ago by Dewey:

> A thing is more significantly what it makes possible than what it immediately is. The very conception of cognitive meaning, intellectual significance, is that things in their immediacy are subordinated to what they portend and give evidence of. An intellectual sign denotes that a thing is not taken immediately but is referred to something that may come in consequence of it.[4]

The meaning of any object, quality, event, or action is what it points to by way of some experience. Meaning is relational, and the meaning of a certain object would be the possible experiences it affords us—either now, in the past, or in the future (as possibilities). We can see this relational, experiential, enactive character of meaning in our understanding of physical objects and events. Psychologist Lawrence Barsalou coined the term "perceptual symbols" for the sensory-motor-affective representations through which we experience, understand, and think about physical objects in our environment. He argues that "cognition is inherently perceptual, sharing systems with perception at both the cognitive and neural levels."[5] The central idea is that to say that we have a concept of some object is to say that we can neutrally simulate a series of sensory, motor, and affective experiences that would be associated with objects of that sort. Barsalou explains:

> Perceptual symbols are modal and analogical. They are modal because they are represented in the same systems as the perceptual states that produced them. The neural systems that represent color in perception, for example, also represent the colors of objects

> in perceptual symbols, at least to a significant extent. On this view, a common representational system underlies perception and cognition, not independent systems. Because perceptual symbols are modal, they are also analogical. The structure of a perceptual symbol corresponds, at least somewhat, to the perceptual state that produced it.[6]

For example, the meaning of a cup is not just some abstract concept specifying a defining set of features that constitute it as a cup. Rather, the meaning of a cup is all of the experiences, both actual and simulated, it can afford us. Some of those experiences will be mostly present (and sometimes, future) sensory-motor activations, such as the visual properties it presents, the way it feels in my hand as I lift it, the smoothness of its ceramic surface, or its capacity to hold my tea. However, the meaning of the cup is not just what it affords us by way of physical perception and motor interaction, because it also includes the social functions of cups, given our cultural values and practices surrounding the use and significance of various types of cup. Finally, in addition to this public and shared meaning, there will be each individual's own personal past experiences with cups, and perhaps with this very same cup which now sits before him or her. Perhaps it is the cup given to you by your students at the end of a course, which has sat on your desk for the past thirty years, connecting you with those students, that class, and the process of education—not to mention that it has been a good vehicle for holding your morning tea all those years.

In this account of embodied meaning, I have been appropriating J. J. Gibson's notion of perceptual "affordances."[7] Gibson, like Dewey, saw that objects are and mean the possibilities for experience, both actual and simulated, that they afford us. What any object affords is the result of the nature of our bodies and brains—our perceptual apparatus, or neural binding capacities, our affective responses, our motor programs—*as they interactively engage patterns and structures of our environments.* So, for a human being with fingers, hands, and arms, a ceramic cup affords pick-up-ability, whereas for an ant it might provide climb-up-ability. Properties, modes of interaction, and meanings are therefore relative to the character of the organism (live creature) and to the objective characteristics of its environments. The objects that populate our world greet us with their meaningful affordances as we engage them in activity. Such affordances define the spaces in which creatures like us can be "at home" in our world; that is, the affordances define the types of couplings and transformative operations we can experience.

An important component of the new cognitive science of meaning and thought is the utilization of neural imaging research. Each of the experiences related to cups that I described in the previous paragraphs is correlated with activation of functional clusters of neurons. We now know from neural imaging studies that seeing a cup is not just a *visual* experience, but also activates some of the neurons in the motor and premotor cortical areas of the brain that would be activated if we actually picked up the cup, manipulated it, or drank from it.[8] These so-called *canonical* neural clusters are what make our concepts multimodal, involving activation of various modes of perception and interaction connected with a particular object. There are also *mirror neuron* systems, so that when we see someone performing a specific action, many of the same sensory and motor functional clusters of neurons are activated in our brains as if we were performing that action.[9] Although this research began with monkeys, it has been extended to human cognition, and we now know that mirror neuron effects are present even when we imagine performing an action, read about such an action, or hear someone describing the action.[10] This is embodied meaning in the most direct and intimate sense.

The conception of meaning that I have briefly sketched here has been substantially developed by a number of linguists, psychologists, neuroscientists, and philosophers over the past thirty years, especially within what is known as cognitive linguistics. This orientation begins with organism-environment interactions and the mostly nonconscious selection by an organism of certain relations or qualities of these interactions that are taken as signs of other experiences. Barsalou argued that we should think of concepts as "simulators," because they simulate in our brains and bodies the kinds of experiences we might have with particular objects, events, or situations. A similar view has recently been expanded into what is known as simulation semantics. Ben Bergen, a cognitive scientist with a background in linguistics, describes the *embodied simulation hypothesis* as the view that "we understand language by simulating in our minds what it would be like to experience the things that the language describes."[11] We use the same systems for conceptualization, reasoning, and communication that we use in experiencing the objects we conceptualize, reason about, and write and talk about.

The idea is that you make meaning by creating experiences for yourself that—if you are successful—reflect the experiences that the speaker, or in this case the writer, intended to describe. Meaning, according to the embodied simulation hypothesis, is not just abstract mental symbols; it is a creative process, in which people construct virtual experiences—embodied simulations—in their brains.[12]

Bergen's focus is primarily on the dynamic, constructive processes of meaning-making in language. However, I want to suggest that the enactive, simulative processes of meaning and thought he is describing are equally present in experiences of meaning that are not merely linguistic. Any form of symbolic interaction will manifest these same sensory-motor-affective simulations that are meaningful to humans, given their bodily makeup, the types of environments in which they dwell, and the cultural history, practices, and institutions they inhabit. The crux of this view is that meaning is not just some abstract, disembodied conceptual content, but rather involves the neural simulation of sensory, motor, and affective processes that we associate with the thing or event that has meaning for us. My point is that the only way to explain the meaning and power of architectural affordances in our lives requires these multimodal, enactive, simulation processes of meaning-making.

THE QUALITATIVE DIMENSIONS OF EXPERIENCE AND MEANING

In addition to the previously sketched account of meaning, the second thing I want to bring to bear on our understanding of how architecture is meaningful is the central role of the qualitative aspects of experience. We dwell in a world of qualities—the fresh earthy scent of a cool breeze coming in through the window on a spring morning, the sounds of children playing, the honking of horns in congested traffic accompanied by the smell of exhaust and the feeling of cars and trucks pressing in around us, and the refreshing shock of the cold mountain lake after a strenuous sweaty hike. We act to realize some qualities and avoid others. Our Eros draws us to the eyes of our lover and to their scent, skin, breath, and lips—all of which are experienced qualitatively without any need for reflective thought. Our world is a realm of immediately felt qualities that have meaning for us even before and without language. This is not to deny that language and other forms of symbolic interaction can sometimes dramatically enrich our possibilities for meaning, but linguistic meaning is already itself parasitic on embodied, qualitative meaning.

Dewey argued at great length that, although the sciences are important because of the way they selectively abstract out certain relations among objects and events and then use those relations to frame causal laws, it is precisely such abstractions that ignore the qualitative dimensions of our experience, in favor of causal connections. What is needed to reconnect our scientific understandings to our lived experience is recognition of the role of qualities in what and how things are meaningful to us.

In this context, one of Dewey's most radical ideas was that, in addition to specific sensory qualities, every situation we encounter is unified and marked off by what he called its "pervasive unifying quality":

> An experience has a unity that gives it its name, *that* meal, that storm, that rupture of friendship. The existence of this unity is constituted by a single *quality* that pervades the entire experience in spite of the variation of its constituent parts. This unity is neither emotional, practical, nor intellectual, for these terms name distinctions that reflection can make within it.[13]

To find yourself enmeshed in an experience is to feel the qualitative unity that gives meaning and identity to what is happening to you. Dewey's argument is that only within such a unified situation do we *then* experience individual objects, persons, and events, with their particular qualities and affordances. Humans, like all other animals, are selective creatures; that is, our survival and flourishing depend on our ongoing, mostly unconscious, selection of aspects of our environment for attention, interaction, and transformation. So, *objects are events with meanings* that "stand out" within the context of a situation. An object is "some element in the complex whole that is defined in abstraction from the whole of which it is a distinction. The special point made is that the selective determination and relation of objects in thought is controlled by reference to a situation—to that which is constituted by a pervasive and internally integrating quality."[14] Objects, then, are clusters of affordances of possible interactions we have had, or might have. Objects stand out for us because they are significant for the kinds of creatures we are, with the kinds of perceptual and motor capacities we have, and the kinds of purposes and values we cherish: "things, objects, are only focal points of a here and now in a whole that stretches out indefinitely. This is the qualitative 'background' which is defined and made definitely conscious in particular objects and specified properties and qualities."[15]

The relevance for architecture of this conception of the qualitative unity of a situation or experience is that any encounter with an architectural structure begins with a felt qualitative sense of our whole situation, prior to any definite attention to component parts, relations, or qualities. Dewey's way of making a similar point (though one not explicitly directed to architecture) is to differentiate the *sense* of a situation from the *signifying* functions of various elements within that situation.

> The qualities of situations in which organisms and surrounding conditions interact, when discriminated, make sense. Sense is distinct from feeling, for it has a recognized reference; it is the qualitative characteristic of something, not just a submerged unidentified quality or tone. Sense is also different from signification. The latter involves use of a quality as a

> sign or index of something else. ... The sense of a thing, on the other hand, is an immediate and immanent meaning, it is meaning which is itself felt or directly had. ... The meaning of the whole situation as apprehended is sense.[16]

So, the sense of a situation or experience is felt as a meaningful whole within which we then discriminate relevant objects, qualities, events, and persons that matter to us because they signify—point to—other objects, events, relations, and experiential consequences that are connected to our present situation. *My hypothesis is that architectural structures are experienced by humans as both sense-giving and signifying. That is, architectural structures present us, first, with a way of situating ourselves in, or being "at home" in, and making sense of our world, and, second, they provide material and cultural affordances that are meaningful for our survival and flourishing as meaning-seeking creatures.* Consequently, any encounter with an architectural structure begins with the overall sense of place (of being in a particular world), followed almost immediately by a growing grasp of the numerous meanings afforded by its various parts, light patterns, structural relations, contrasts, flow, rhythms, and other significant elements of meaning within the work.

BODILY STRUCTURES OF MEANING IN ARCHITECTURE

I am suggesting that we need an embodied view of mind and meaning to appreciate the significance of architecture. Any architectural theory based on a disembodied view will therefore be proportionately inadequate to the extent that it overlooks the embodiment of meaning. Examples of this shortcoming are certain types of computational modeling that disregard history and embodied consciousness. Alberto Pérez-Gómez observes that computational architectural models cannot incorporate the prereflective and embodied dimensions and qualities that ground human meaning:

> These instrumental processes are necessarily dependent on mathematical models, and often become an empty exercise in formal acrobatics. Architects soon forget the importance of our verticality (our spatial engagement with the world that defines our humanity, including our capacity for thought), our historicity (we are, effectively, what we have been), and gravity (the "real world" of bodily experience into which we are born, and that includes our sensuous bond to all that which is not human).[17]

Pérez-Gómez is criticizing misguided attempts to model our sense of space and place in a manner that presupposes the mistaken view that mind is disembodied and that human understanding can be decontextualized. If we forget our embodiment, and the fact of our being situated within a particular concretely experienced environment, we lose the very means for explaining the power and importance of architecture. Thus, Dalibor Vesely concludes that "to perceive, to move and to learn, in the human world is possible only due to a corporeal involvement. The disembodied nature of computer programs is the main reason for their inability to match human intelligence."[18]

What we need, then, is an understanding of "mind" as embodied and enactive relative to our experience and comprehension of architectural spaces. I am arguing that such an account needs to keep in mind two fundamental points: (1) that the meaning of any object is grounded in the affordances for possible experiences related to that object; and (2) our account of these affordances must include the crucial role of the qualitative dimensions of any experience, especially the pervasive unifying quality of the situation. I want to put some flesh on these skeletal claims by giving a few brief examples of some of the more important patterns of our embodied interactions that thereby gain considerable significance for our experience of architecture.[19] In *The Body in the Mind* and *The Meaning of the Body*, I described various *image-schematic* patterns of recurring structures of experiences that humans (and some other animals) encounter through our mostly unreflective bodily engagement with our environments. I argued that such image schemas are immediately significant for us through the affordances they provide for how we can meaningfully interact with our world. Body-based meaning structures of this sort have obvious relevance for the meanings of constructed environments. Here are a few of the more important image-schemas, with some indication of their relevance for architecture.

CONTAINMENT

Life plays out within boundaries. Our bodily organisms are defined by semipermeable boundaries into which we must incorporate energy and out of which we must expel waste. As Antonio Damasio argues, life goes on within the boundaries that define organisms, and "for whole organisms, then, the primitive of values is *the physiological state of living tissue within a survivable, homeostatic range*."[20] Whatever else we do, our bodies must maintain an appropriate dynamic equilibrium internally, in ongoing response to changing aspects of our environments. Otherwise, we become dysfunctional, and may even perish.

We thus learn the meanings of containment in the most intimate bodily way, first through our visceral sense of our bodies as containers, and then through our bodily manipulations of containers. Very early on, babies begin to learn the "spatial logic" of containment, as they play with nesting objects (cups, bowls, boxes) and experience, through their bodily interactions, movements into and out of bounded spaces. They learn that if the small ball is placed in cup A, and cup A is nested within cup B, then the ball is "in" cup B. In formal logic, this is known as the transitivity relation (If A is in B, and B is in C, then A is in C), but babies learn this logical relationship as a spatial or corporeal logic not reflectively, but in a bodily fashion through sensory-motor activities.

This kind of *ecological logic* lies at the heart of our experience of architecture, so that we learn the meaningful affordances of particular kinds of containment structures, in relation to our bodily makeup, needs, desires, and ideals. Consider, for instance, the sheltering function of much architecture. Shelter requires a relative strength, stability, and at least some measure of impenetrability. We learn which materials are strong, which insulate, and which are available and cost-effective, and we respond in certain ways emotionally to structures that we feel to be strong, solid, and well-grounded. Typically, we also desire to be enclosed in spaces that are not claustrophobic and oppressive. From infancy on we climb in and out of boxes, baskets, cribs, closets, cars, and other containers. We find what it feels like to be confined within tight containers, as compared to roaming more freely in open spaces. We know how bad it can feel to be "boxed in." There is a way it feels to be confined in a relatively closed, dark, damp space (e.g., a cave), which is experientially quite different from flat, open, sweeping expanses of the plains, or from standing high on some mountain with an elevated view of the world spread out below you. Consequently, we come to desire shelters of a certain size, height, and configuration, depending on our purposes. Most of us do not feel at home in completely closed containers. We want access to light and air, so we want windows and doors that open us to exchanges with our surroundings. We want a certain, perhaps culturally variable, degree of privacy, but we need ways to learn what is going on in the world beyond our door or gate. In other words, we want to be in and of the world when we think that serves our purposes, and we want shelter and privacy at other times.

VERTICALITY (AND OTHER SPATIAL ORIENTATIONS)

Insofar as we are creatures embraced by gravity, what goes up must come down. This is something we have to learn as children, but no reflection is required for this, since we only have to observe how objects and people move in their environment. We therefore dwell at least partly in an up-down world. Because of gravity, the very accomplishment of rising up requires effort, power, control, and balance. One of the most significant human transitions from infancy to childhood is the emergence of an upright posture. We struggle to stand erect, and we learn that standing requires a firm base (ground) and an appropriate balance.

Our mundane encounters with the meanings of verticality give rise to distinctive architectural experiences. Contrast, for example, the experience in a Gothic cathedral of being carried upward with your gaze into an unknowable darkness beyond this world versus the Hopi focus on the kiva as the portal from which spirit enters our world out of the earth or ground. The Christian cathedral is meant to direct us upward toward a projected supernatural realm of perfect and complete Being, while the kiva locates us on the earth from which life, and the universe, supposedly spring.

BALANCE

The maintenance of balance is one of the key values of all living organisms. We must maintain both internal balance (homeostasis or allostasis), and we strive for bodily balance as the basis for our capacity to remain upright and in control. Balance is first encountered by us as a bodily experience of our shaky relation to our surroundings. Babies, after many trials over an extended period, come to tentatively master an upright posture. We are aware of the crucial role of balance in our lives mostly when we lose it, rather than when we unconsciously achieve and maintain it. We eventually learn to project the qualities of our felt experience of balance onto objects that populate our world. The Leaning Tower of Pisa makes some people feel vertiginous and others slightly uneasy. Some experience it as profoundly unbalanced and disturbing. Likewise, Richard Serra's *Tilted Arc* made many people so uncomfortable that it was removed eight years after its installation in the Federal Plaza in Manhattan. Some complained that the sculpture partitioned the public space in a way that dramatically restricted public access and movement in the plaza, and others felt threatened by the imposing 12-foot-high, 120-foot-long steel wall tilting precariously over them. For them, this did not create a happy space where they could feel comfortable in their comings and goings.

One of Dewey's more controversial claims was that the feelings and emotions that Western cultures tend to attribute to the subjective inner states of persons ought actually to be recognized as defining the objective situation. Instead of saying merely "*I* am fearful," we ought rather to say, "The situation is fearful." What Dewey saw was that the proper locus of the affective is the entire cycle of organism-environment interaction, and not just the internal states of the organism. Dewey's view gives us a way of making sense of that fact that some buildings can be unbalanced, teetering on collapse. The *building*—the physical object—is unbalanced, and this is not just a way of talking about the feelings we have when we see and interact with it.

FORCES

Our world is a scene of ongoing forceful interactions of energy fields. From infancy, we are lifted, lowered, turned over, patted, stroked, squeezed, buffeted about, breathed on, constrained, contained, ministered to, fed, burped, wiped, rocked, comforted, kissed, talked to, and on and on through all of the forceful bodily events that make up our surroundings. In this intimate bodily way, we learn the types, consequences, and meanings of the various forces within our cosmos.

Physical structures forcefully shape the range of actions possible for us in our environment. You may enter here, but not there. You must walk up these steps, or down this stairway, to gain access. You may, or may not, open this door or window. You must move along this narrow corridor. You may or may not tarry here in this space. All of these experiences of restricted or free access involve structured forceful interactions. Even when we merely see a building, before ever entering it, we *feel* its affordances for how it will forcefully shape our engagement with it.

Massive tilting objects tend to frighten us, because we have learned that forces of nature and gravity tend to topple such tilting objects, unless they are very strong and firmly rooted. Large, heavy objects supported by thin legs seem unstable to us. The fall of a building is a powerful event that reaches deep down into our emotional experience of falling, disintegration, and the release of forces. We are mesmerized by building implosions and demolitions, which we experience as almost sublime—all of that energy released as all twenty stories crumble to the ground, sending out a shockwave of dust. We are overwhelmed by the catastrophic destabilization that occurs as massive forces are released and the structural integrity fails. The collapse of the World Trade Center towers bore tragic witness to this horrific experience.

MOTION

A vast amount of the information one receives about the world comes as a result of our ability to move ourselves within our environment and to move our hands over surfaces. At a very deep level, we learn the contours of our world and the possible ways we can interact with it via movement. As Maxine Sheets-Johnstone has argued:

> In the beginning, we are simply infused with movement—not merely with a *propensity* to move, but with the real thing. This primal animateness, this original kinetic spontaneity that infuses our being and defines our aliveness, is our point of departure for living in the world and making sense of it.[21]

Even though most of the time we are not consciously aware of our bodily movements, we continually experience the qualities of different types of movement. We feel the rhythms of various movements—short, jerky hop-and-skip motions versus smooth, continuous flowing motions. We contrast, within our bodies, the felt difference of gradual accelerations and decelerations versus jolting starts and finishes. Felt rhythms provide basic types of contours for our experience. Events speed up and slow down, creep along, rush past, dance, stumble, drag by, and float. By complex perceptual and cognitive processes, we learn to experience what we might call perceptual motion in physically fixed or static visual arrangements. A series of connected Romanesque arches carry our perception along in a smooth, recurring pattern of curving visual motion. The jarring angles of certain Kandinsky paintings have a very different felt quality of perceptual motion, in contrast, say, with the curvilinear, organic motions of a Matisse landscape.

This same type of perceptual motion experience plays a key role in our experience of buildings. For example, the flowing, playful, and sometimes incongruous angles and lines of various postmodern designs present a very different overall unifying quality than the austere, machine-like regularities and rectilinearities of modernist glass-wall box structures. Some will prefer Mies van der Rohe, others Art Moderne lines, others Gaudí's organic ecologies, and still others Gehry's playful postmodernism, because of the way each of these markedly different qualitative unities affords us dramatically different imaginative experiences for how we can engage and interact with those structures.

What I have been suggesting, via this brief reflection on image-schematic structures in architecture, is that architecture can provide us with meaning in at least two different, but related, ways. First, every architectural structure will present us with a felt qualitative unity of the whole that, in essence, gives us a world (however small), and a certain way of inhabiting that world. Second, the building's particular affordances provide the possibilities for meaningful engagement with the building or constructed space, in relation to its particular structures and qualities. We can talk about these meanings using language, but the meanings are not, as we have seen, for the most part linguistic in nature. Rather, they employ the meaning of our bodily interactions with our environments, and this exists prior to and beneath our linguistic resources.

At the level of the overall qualitative unity of the work, it can be said that each building gives us a world that we can inhabit—not just a physical world, but a social and cultural world with its defining values. With respect to narrative, Paul Ricoeur was fond of claiming that each narrative work (historical or fictional) offers us a "world of the work" into which we can project our inmost needs and desires for meaning and value.[22] Going beyond narrative to include architecture, Heidegger claims that the Greek temple enacts a world of the fourth- or fifth-century BCE Athenian: "The temple-work, standing there, opens up a world and at the same time sets this world back again on earth, which itself only thus emerges as native ground."[23] Although neither Ricoeur nor Heidegger uses Dewey's language of the pervasive unifying quality of a situation or of an experience, Dewey's description of our felt encounter with a loosely or tightly ordered world applies to their examples. It is for this reason that we are prone to identify particular works as embodying historical and cultural experiences and values that capture certain salient characteristics of a specific time, place, and social reality.

FROM FITNESS TO FLOURISHING

In describing but a very few of the image-schematic patterns that are immediately meaningful to us via our embodiment, we have already moved across the boundary between fitness and flourishing. We have made the transformation from what we value for the sake of our survival and fitness to what we value because we crave enriched and deepened meaning. This exposes the artificiality of the fitness/flourishing distinction, because both fitness and flourishing are about how things are meaningful and significant for the kinds of bodily and cultural creatures we are in relation to our physical and cultural environments. In this way, we have slipped almost imperceptibly from architecture's functional usefulness for survival to its capacity to give us meaning and to present ideals for how our world might be transformed.

Architecture is thus one of our most human and potentially humane ways of relating to our environing world. What Dewey says about all art—that it is a form of human meaning-making—is amply illustrated by architecture. Art takes the physical and cultural materials of our embodied and social experience and transforms them into new experiences that intensify, harmonize, and enrich meanings and possibilities for living and acting in the world. Contrary to many traditional theories that isolate "fine" art from ordinary life, Dewey recognized that art brings to consummation and fulfillment materials and aesthetic dimensions that permeate our everyday experiences. Dewey saw art not as disengaged from, or rising above, ordinary life, but rather as continuous with other forms of making and problem-solving that intelligent creatures utilize to improve the quality of their lives.

Architecture is a wonderful example of this meaning-making process. It grows out of our need for shelter and a more or less harmonious relation to our surroundings. It is a response to the human problem of dwelling safely and happily in our world. It draws on our ability to fabricate structures and to transform the materials we find in nature. It is a form of problem-solving that equally addresses our need for physical security and our need for meaning and aesthetic well-being (where aesthetics is about everything that goes into our capacity to have any sort of meaningful experience).[24] In short, architecture is an act of imaginative problem-solving and meaning-making that resonates with the deepest levels of our connection to our environment. In the words of Juhani Pallasmaa, "architecture is a mode of existential and metaphysical philosophy through the means of space, matter, gravity, scale and light."[25] This captures what I have been saying about the embodied and existential way architecture explores the possible ontologies and ecologies of the human world, though without the use of linguistic propositions that are the darlings of linguistically articulated philosophical systems. Instead, architecture helps us enact what George Lakoff and I have called a "philosophy in the flesh."[26] Drawing on the work of the phenomenologist Maurice Merleau-Ponty, Pallasmaa emphasizes architecture's ability to realize various ways of being in, or inhabiting, our world, in the most intimate, embodied, situated manner:

> The task of architecture is "to make visible how the world touches us," as Maurice Merleau-Ponty wrote of the painting of Paul Cézanne. … We live in the "flesh of the world," and architecture structures and articulates this existential flesh, giving it specific meanings. Architecture tames and domesticates the space and time of the flesh of the world for human habitation. Architecture frames human existence in specific ways and defines a basic horizon of understanding.[27]

The term "understanding" in this passage refers not to some conceptual or propositional belief structure, but rather to a certain specific way of inhabiting and being at home in one's world. I have described but a few of the modes of embodied meaning by which this existential sense-giving plays out in our constructed spaces. My contribution to this discussion has been primarily the elaboration of a theory of embodied meaning adequate to the task of understanding how architecture enacts and transforms meaning, including the role of the qualitative dimensions of meaningful experience. I have emphasized that the meaning involved here is not so much symbolic meaning, but rather a more direct, embodied realization of sense and quality and significance for our lives. I think Pallasmaa captures this characterization of meaning as embodied, qualitative, and affective when he says: "architecture mediates and evokes existential feelings and sensations. The buildings of Michelangelo, for instance, represent an architecture of melancholy and sorrow. But his buildings are not symbols of melancholy, they actually mourn."[28] This is a prime example of what I am calling "the embodied meaning of architecture."

I want to close by reiterating my claim that buildings do not merely reveal our sense of our world and manifest our embodied ways of making sense of that world. If architecture had only this representative function, it might not do anything more than merely express some person's or society's dysfunctional, inharmonious, and highly problematic situation. Indeed, this is often what bad architecture does—it holds before us and habituates the impoverishment of our lives, social arrangements, and relations to our environment and to other people.

To state what most will consider obvious, architecture at its best goes beyond the mere expression of a world to creatively transform the conditions of our human habitation and interaction. This is its moral imperative—to make the world a better place in which to live. It carries out this task whenever it helps resolve the problematic situations in which people find themselves, and when it enriches the meaning and growth of our experiences. Because human meanings and values are plural and complex, there can be no single universal way to realize "the better" for human existence.[29] Nevertheless, even as we must embrace a certain pluralistic set of architectural norms for growth and enrichment of meaning, this does not mean that we cannot, in a particular context, determine better and worse solutions to our need for meaningfully ordered spaces and buildings, as anyone knows who has ever lived in a crummy apartment.

NOTES

Approximately one-third of this chapter is based on an earlier article published as "Architecture and the Embodied Mind," *OASE* 58 (2002): 75–96, though with considerable changes made to the earlier text, and substantial additions of new material.

1. Owen Flanagan, *The Really Hard Problem: Meaning in a Material World* (Cambridge, MA: MIT Press, 2007); Thomas M. Alexander, *The Human Eros: Eco-ontology and the Aesthetics of Existence* (New York: Fordham University Press, 2013).

2. John Dewey, *Art as Experience* (1934), vol. 10 of *The Later Works, 1925–1953*, ed. Jo Ann Boydston (Carbondale: Southern Illinois University Press, 1987), 13.

3. In fact, the view of meaning as conceptual/propositional/linguistic turns out to be inadequate in explaining even how sentences have meaning for us. George Lakoff and Mark Johnson, *Philosophy in the Flesh: The Embodied Mind and Its Challenge to Western Thought* (New York: Basic Books, 1990); Mark Johnson, *The Meaning of the Body: Aesthetics of Human Understanding* (Chicago: University of Chicago Press, 2007).

4. John Dewey, *Experience and Nature* (1925), vol. 1 of *The Later Works, 1925–1953*, 105.

5. Lawrence Barsalou, "Perceptual Symbol Systems," *Behavioral and Brain Sciences* 22 (1999): 577.

6. Ibid, 578.

7. James J. Gibson, *An Ecological Approach to Visual Perception* (Boston: Houghton Mifflin, 1979).

8. Vittorio Gallese and George Lakoff, "The Brain's Concepts: The Role of the Sensory-Motor System in Conceptual Knowledge," *Cognitive Neuropsychology* 21 (2005): 1–25.

9. Giacomo Rizzolatti and Laila Craighero, "The Mirror-Neuron System," *Annual Review of Neuroscience* 27 (2004): 169–192.

10. Vittorio Gallese and Alvin Goldman, "Mirror Neurons and the Simulation Theory of Mind-Reading," *Trends in Cognitive Science* 2 (1998): 455–479; Jerome Feldman, *From Molecule to Metaphor: A Neural Theory of Language* (Cambridge, MA: MIT Press, 2006); Benjamin Bergen, *Louder Than Words: The New Science of How the Mind Makes Meaning* (New York: Basic Books, 2012).

11. Bergen, *Louder Than Words*, 13.

12. Ibid, 16.

13. Dewey, *Art as Experience*, 37.

14. John Dewey, "Qualitative Thought," in vol. 5 of *The Later Works, 1925–1953*, 242–262 (246).

15. Dewey, *Art as Experience*, 197.

16. Dewey, *Experience and Nature*, 200.

17. Alberto Pérez-Gómez, "Phenomenology and Virtual Space: Alternative Tactics for Architectural Practice," *OASE* 58 (2002): 36.

18. Dalibor Vesely, "Space, Simulation and Disembodiment in Contemporary Architecture," *OASE* 58 (2002): 66.

19. I am painfully aware of the shortcomings of this present section. I trust it is clear that I am here only able to make somewhat sweeping gestures toward but a few of the many embodied structures of meaning that are relevant to all experience, and hence to all architecture. I make no pretense to show how these operate concretely for any particular building. An adequate argument would at least require a fairly detailed treatment of the working of various embodied aspects of meaning to generate our complex experience of the meaning and power of a particular architectural space and structure.

20. Antonio Damasio, *Self Comes to Mind: Constructing the Conscious Brain* (New York: Pantheon, 2010), 49.

21. Maxine Sheets-Johnstone, *The Primacy of Movement* (Amsterdam: John Benjamins, 1999), 136.

22. Paul Ricoeur, *Time and Narrative*, vol. 1 (Chicago: University of Chicago Press, 1983).

23. Martin Heidegger, "The Origin of the Work of Art," in *Poetry, Language, Thought*, trans. Albert Hofstadter (New York: Harper and Row, 1971), 42.

24. Johnson, *The Meaning of the Body*, 209ff.

25. Juhani Pallasmaa, "Lived Space, Embodied Experience and Sensory Thought," *OASE* 58 (2002): 26–28.

26. Lakoff and Johnson, *Philosophy in the Flesh*, 551–568.

27. Pallasmaa, "Lived Space, Embodied Experience and Sensory Thought," 18.

28. Ibid., 22.

29. Mark Johnson, *Morality for Humans: Moral Understanding from the Perspective of Cognitive Science* (Chicago: University of Chicago Press, 2014).

3

BODY, MIND, AND IMAGINATION: THE MENTAL ESSENCE OF ARCHITECTURE

Juhani Pallasmaa

If the body had been easier to understand, nobody would have thought that we had a mind.[1]

Richard Rorty

Instead of stepping on the specialized ground of neuroscience, I wish to elaborate on the specific mental essence of architecture—a realm that is deeply biologically and culturally grounded, although poorly understood in both education and practice. It is my hope that the exciting doors that the biological and neurosciences are now opening will valorize the interaction of architecture and the human mind, and reveal hidden complexities that have thus far escaped measurement and rational analyses. In our postmodern society, dominated by shallow rationality and reliance on the empirical, measurable, and demonstrable, the embodied and mental dimensions of human existence are continually suppressed. I believe that neuroscience can lend support to the mental objectives in design and the arts, which are in danger of being eliminated because of their "practical" uselessness and apparent subjectivity. Architecture has its utilitarian qualities in the realm of rationality and measurability, but its mental value is most often concealed in embodied metaphors and ineffable unconscious interactions—it can only be experienced and encountered. As Jean-Paul Sartre argues, "Essences and facts are incommensurable, and one who begins his inquiry with facts will never arrive at essences ... understanding is not a quality coming to human reality from the outside; it is its characteristic way of existing."[2]

Rather than attempting to highlight the new insights of neuroscience that could be applicable to architecture, I have chosen to focus on the mental dimensions of building that could be valorized by new scientific research. I believe that neuroscience can reveal and reinforce the fundamentally mental, embodied, and biological essence of profound architecture against current tendencies toward increasing materialism, intellectualization, and commodification. I will attempt to illustrate the mental and spiritual qualities of architecture and art side by side, as I see the craft of architecture, in its existential and mental dimensions, also as an art form. No doubt, architecture is ontologically grounded in utility and technological reality, and this makes it equally decisively a nonart. While writing this essay at Taliesin West, Frank Lloyd Wright's studio in the Arizonan desert, every morning I saw a quote of Frank Lloyd Wright printed on my tea mug: "I believe a house is more a home by being a work of art."[3] For my purposes in this context, architecture is and is not an art, depending on one's point of view.

THE TASK OF ARCHITECTURE

The purpose of our buildings is too often understood solely in terms of functional performance, physical comfort, economy, symbolic representation and aesthetic values. However, the task of architecture extends beyond its material, functional, and measurable properties—and even beyond aesthetics—into the mental and existential sphere of life. Buildings do not merely provide physical shelter or facilitate distinct activities. In addition to housing our fragile bodies and actions, they must also house our minds, memories, desires and dreams. Buildings mediate between the world and our consciousness through internalizing the world and externalizing the mind. Structuring and articulating lived existential space and situations of life, architecture constitutes our most important system of externalized order, hierarchy, and memory.

We know and remember who we are as historical beings by means of our constructed settings. Architecture also concretizes "human institutions," to use a notion of Louis Kahn's, the layering of cultural structures, as well as the course of time. It is not generally acknowledged that our constructed world also domesticates and scales time for human understanding. Yet architecture slows down, halts, reverses, or speeds up experiential time, and we can appropriately speak of slow and fast architectures. As the philosopher Karsten Harries suggests, architecture is "a defense against the terror of time."[4] It gives limitless and meaningless space its human measures and meanings, but it also scales endless time down to the limits of human experience; the mere memorized image of the Egyptian pyramids concretizes the distance of four thousand years in our consciousness. It is evident that architecture has the tendency to turn ever faster in our era of speed and acceleration. Finally, Gaston Bachelard assigns a truly monumental task to architecture: the house "is an instrument with which to confront the cosmos."[5] He criticizes the Heideggerian assumption of the basic human frustration arising from "being cast into the world," as, in his view, we are born "in the cradle of architecture,"[6]

3.1 In addition to "domesticating" physical space for human use and grasp, architecture "tames" time for human understanding. The Great Pyramids of Gizeh.

not cast into meaningless space. Indeed, until the Renaissance, the main mental task of architecture was to mediate between macrocosm and microcosm, the divinities and the mortals. "With the Renaissance revival of the Greek mathematical interpretation of God and the world, and invigorated by the Christian belief that Man as the image of God embodied the harmonies of the Universe, the Vitruvian human figure inscribed in a square and a circle became a symbol of the mathematical sympathy between microcosm and macrocosm," Rudolf Wittkower informs us.[7] Today, architecture has become mere utility, technology and visual aesthetics, and we can sadly conclude that it has abandoned its fundamental metaphysical task.

The human essence of architecture cannot be grasped at all unless we acknowledge its metaphoric, mental, and expressive nature. "Architecture is constructed mental space," Finnish professor Keijo Petäjä used to say.[8] In the Finnish language, this formulation projects two meanings simultaneously: architecture is a materialized expression of human mental space; and our mental space is itself structured and extended by architecture. This idea of a dialectical relationship, or interpenetration of physical and mental space, echoes Maurice Merleau-Ponty's phenomenological notion "the chiasmatic bind"[9] of the world and physical space, on the one hand, and the self and mental space, on the other. In his view this relationship is a continuum, not a polarity. The chiasmatic continuum of outer physical and inner psychic space can, perhaps, be illustrated by the enigmatic image of the Moebius strip, a looping ring that has only one continuous surface. It is exactly this

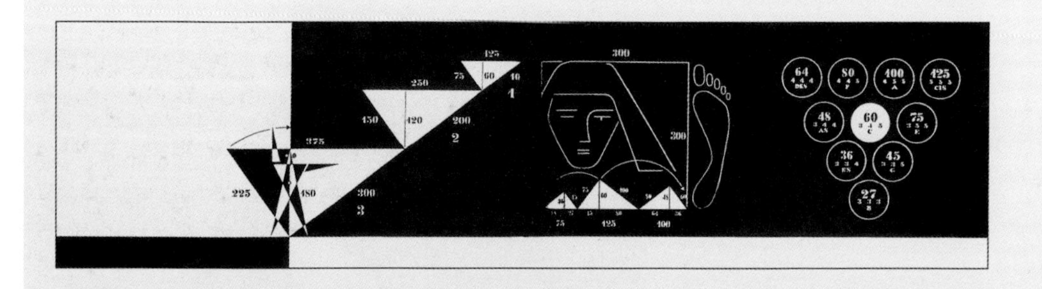

3.2 Pythagorean studies in mathematically based harmony were revitalized during the Renaissance, and again during the twentieth century. The aim of Pythagoreanism is to create a shared harmonic ground for visual phenomena and music. Aulis Blomstedt, Canon 60, around 1960. Professor Blomstedt's system of measures and proportions is based on ten basic numbers and their musical equivalents.

chiasmatic merging and mirroring of the material and the mental that has rendered artistic and architectural phenomena immune to an empirical scientific approach; the artistic meaning exists in the experience of the material realm, and this experience is always unique, situational, and individual. Artistic meaning exists only on the poetic level in our direct encounter with the work, and it is existential rather than ideational—emotional rather than intellectual. Merleau-Ponty also introduced the suggestive notion of "the flesh of the world,"[10] the continuum of the world, which we are bound to share with our bodies as well as with our architecture. In fact, we can think of works of architecture as specific articulations of this very existential and experiential flesh.

FORMAL AND EXPERIENTIAL SPACE

Like most architects my age, I was educated to regard our craft primarily as the construction of visual and aestheticized spatial structures possessing distinct formal characteristics and qualities. Gradually, I have learned to confront buildings experientially as encounters between physical structures and my own existential sense through internalizing multisensory perception. This encounter turns physical and geometric space into existential and lived space, and I become myself an ingredient and measure of the experience itself. This understanding puts the experiencing individual in the very center of the experience. In my way of thinking, a sincere architect cannot authentically design a house facing the client as an external other; the architect has to internalize the client, to

3.3 Juhana Blomstedt (painter son of architect Aulis Blomstedt), *Model and the Artist* (1049) from the Moebius Series, 2003. Oil on canvas, 50 × 50 cm. Private collection.

turn himself into the client, and eventually design the building for him/herself. At the end of the design process, the architect offers the house to the real dweller as a gift. Profound architecture is always a gift of imagination, as it necessarily transcends its given points of departure and factual conditions. It is always bound to contain qualities that no one could have expected or foreseen. This process is similar to the gift a woman makes when she offers her womb to give life to a child on behalf of a woman who is physiologically unable to bear one. Architecture is born of imaginative empathy, and the talent of compassion is as important to the architect as formal fantasy.

BOUNDARIES OF THE SELF

"What else could a poet or painter express than his encounter with the world?" Merleau-Ponty asks.[11] An architect is bound to articulate this very same personal encounter, regardless of the basic utility and rationality of his/her task and the fact that he/she is engaged primarily in creating settings for others. This might sound like a self-centered position, but in fact, it emphasizes and concretizes the subtlety of the designer's human responsibility. In an essay written in memory of Herbert Read, Salman Rushdie suggests: "Literature is made at the boundary between self and the world, and during the creative act this borderline softens, turns penetrable and allows the world to flow into the artist and the artist to flow into the world."[12] Profound works of architecture also sensitize the

3.4 Balthus (Balthazar Klossowski de Rola), *Les Beaux Jours* (The Happy Days), 1944–1946. Oil on canvas, 148 × 200 cm. Hirshhorn Museum and Sculpture Garden, Smithsonian Institution, Washington.

boundary between the world and ourselves; I experience this moment and my relationship with the world in a deep and meaningful manner. The architectural context gives my experience of being its unique structure and meaning through projecting specific frames and horizons for my perception and understanding of my own existential situation. The poetic experience brings me to a borderline—the boundary of my perception and understanding of self—and this encounter projects a sense of existential meaningfulness.

SELF-EXPRESSION AND ANONYMITY

Particularly in today's artistic world that seeks novelty and effect, the arts and architecture are seen as modes of the artist's and architect's self-expression. I have become increasingly doubtful about this attitude. Balthus (Balthazar Klossowski de Rola), one of the greatest figurative painters of the twentieth century, is critical of the idea of artistic self-expression: "If a work only expresses the person who created it, it was not worth doing. ... Expressing the world, understanding it, that is what seems interesting to me."[13] This is a rather unexpected attitude from an apparently very self-absorbed painter. He goes even further to demand a distinct anonymity in artistic works: "Great painting has to have universal meaning. This is really no longer so today and this is why I want to give painting back its lost universality and anonymity, because the more anonymous painting is, the more real it is."[14] Again, I suggest that the same criterion applies to the field of architecture, but this is certainly an unfashionable view in today's world obsessed with formal uniqueness and global star architecture.

UNITING THE OPPOSITES

Merleau-Ponty formulates the idea of the world as the primary subject matter of art (and architecture, we might add) as follows: "We come to see not the work, but the world according to the work."[15] As we come to see Frank Lloyd Wright's Taliesin West, we

3.5 Frank Lloyd Wright's Taliesin West is simultaneously an integral part of the Arizonan desert landscape and its geometric and tectonic counterpole. Architecture underlines the landscape and heightens its character. Frank Lloyd Wright, Taliesin West Studio, Scottsdale, Arizona, 1937–1938.

end up experiencing the landscape, as well as our own sense of existence and self altered, refined, and dignified by the magic of architecture. As we enter this compound, we are placed center stage to experience the desert and the sky, light and shadows, intimacy and vastness, materiality and weightlessness, nearness and distance, in a manner that we have not experienced them before. We are invited inside a unique ambience, an artistically structured world of embodied experiences, which addresses our sense of being, balance, horizon, and temporal duration in a way that bypasses rationality and logic. This architecture seems to have been here forever, exuded by the earth itself like the plants of the desert, but the principles and constituents of this convincing unity seem to be beyond rational and verbal analyses. We simply feel it with the same accuracy that we grasp the nature of a landscape with all its life forms, or "understand" the weather. As Alvar Aalto, the Finnish master and Wright's friend, once wrote: "In every case [of creative work] one must achieve the simultaneous solution of opposites. Nearly every design task involves tens, often hundreds, sometimes thousands of contradictory elements, which are forced into a functional harmony only by man's will. This harmony cannot be achieved by any other means than those of art."[16]

In the case of the settings of Taliesin West, the opposites of caving in and flight, separation and togetherness, enclosure and vista, gravity and weightlessness, visuality and

3.6 Wright's architecture is highly atmospheric and projects a haptic feeling with its varied geometry, formal themes, rhythms, tactile materials, and illumination. Frank Lloyd Wright, Taliesin West Residence, Scottsdale, Arizona, 1937–1938.

hapticity, shadow and softened light, give rise to a superbly orchestrated ensemble of experiences. These experiences seem to have the invigorating richness and unpredictability of natural phenomena, held together by an undefinable artistic cohesion, or atmosphere. This place feels like a primordial ritual setting and a utopian community, a futuristic image and a ruin—all at once. It unites earth and sky, the realms of mortals and divinities. Indeed, architecture is logically an "impure" discipline in its fusion of irreconcilable ingredients, facts and beliefs, quantities and qualities, means and ends.

THE SECRET CODE

The mental content and meaning of an architectural experience is not a given set of facts or elements; it is a unique imaginative reinterpretation and re-creation by each individual. The experienced meanings of architecture are not primarily rational, ideational or verbal meanings, as they arise through one's sense of existence by means of embodied and unconscious projections, identifications and empathy. Architecture articulates and "thickens" our sense of being instead of addressing the domain of rational understanding. The British architect, writer, and educator Sir Colin St. John Wilson illuminates this

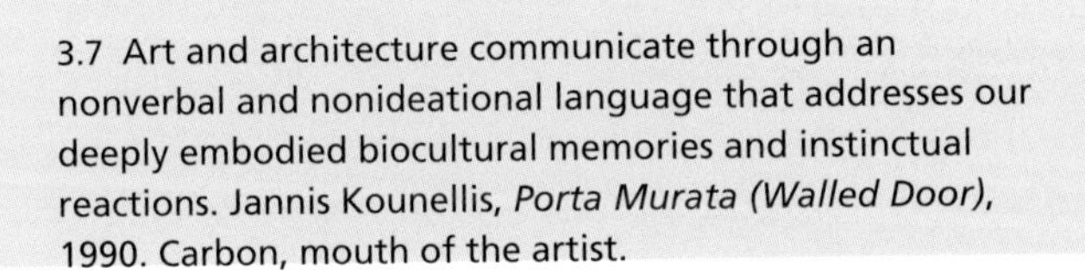

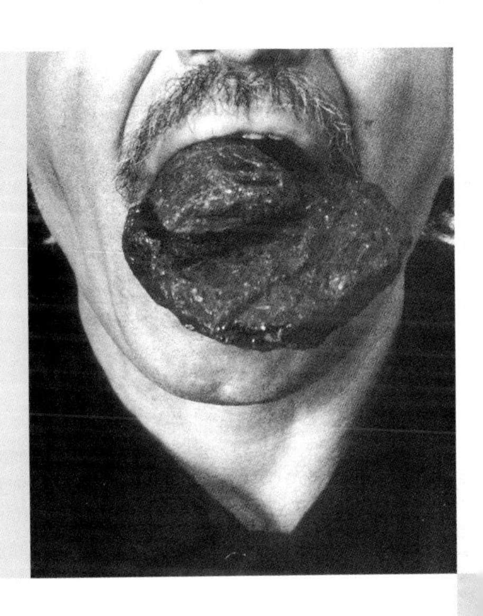

3.7 Art and architecture communicate through an nonverbal and nonideational language that addresses our deeply embodied biocultural memories and instinctual reactions. Jannis Kounellis, *Porta Murata (Walled Door)*, 1990. Carbon, mouth of the artist.

secret, prereflective power of architecture like this: "It is as if I am manipulated by some subliminal code, not to be translated into words, which acts directly on the nervous system and imagination, at the same time stirring intimations of meaning with vivid spatial experience as though they were one thing. It is my belief that the code acts so directly and vividly upon us because it is strangely familiar; it is in fact the first language we ever learned, long before words ... now recalled to us through art, which alone holds the key to revive it."[17]

In his book *The Philosophy of No: A Philosophy of the New Scientific Mind*, written in 1940, Gaston Bachelard describes the historical development of scientific thought as a set of progressively more rationalized transitions from animism through realism, positivism, rationalism and complex rationalism, to dialectical rationalism.[18] "The philosophical evolution of a special piece of scientific knowledge is a movement through all those doctrines in the order indicated," the philosopher argues.[19] In my personal view, profound art and architecture struggle to advance in the reverse direction back toward an animistic relationship with the world, in which we project the world, or we are the world, instead of being outsiders and passive observers. Besides, art is more concerned with our past than with the future. A poetic understanding takes place through unconscious identification, simulation, and internalization. While rational understanding calls for a critical distance and separation from the subject, poetic "understanding" requires nearness, identification, and empathy.

IDENTIFICATION AND EMPATHY

As research has recently revealed, we have a surprising capacity to mirror the behavior of others, and even to unconsciously animate inanimate material constructions and objects. According to Joseph Brodsky, the call of a great poem is to "Be like me."[20] A profound building makes a similar suggestion: "Be a bit more sensitive, perceptive and responsible, experience the world through me." The world of art and architecture is fundamentally an animistic world awakened to life by the projection of our own intuitions and feelings. Paul Valéry regards buildings as entities with distinct voices: "Tell me (since you are so sensitive to the effects of architecture), have you not noticed, in walking about this city, that among the buildings with which it is peopled, certain are *mute*; others *speak*; and others, finally—and they are the most rare—*sing*?"[21] In this sense of searching for an animated and lived world, the artistic intention directly conflicts with science's aim to objectify.

We have an amazing capacity to grasp complex environmental entities through simultaneous multisensory sensing of atmospheres, feelings, and moods. This capacity to instantaneously grasp existential essences of vast entities, such as spaces, places, landscapes and entire cities, suggests that we intuit entities before we identify their parts and details. When discussing the roles of the brain's hemispheres, Iain McGilchrist points out: "The right hemisphere understands the whole not simply as the result of assembling a bunch of fragments, but rather as an entity prior to the existence of the fragments. There is a natural hierarchy of attention, global attention coming first. … You have to see it [an image] as a whole first."[22]

Almost eighty years ago, John Dewey, the visionary pragmatist philosopher, pointed out the significance of such a unifying character and cohesive identity: "An experience has a unity that gives it its name. ... The existence of this unity is constituted by a single *quality* that pervades the entire experience in spite of the variation of its constituent parts. This unity is neither emotional, practical, nor intellectual, for these terms name distinctions that reflection can make within it."[23] "The quality of the whole permeates, affects, and controls every detail," Dewey adds.[24] Sarah Robinson recently pointed out to me a perceptive remark of Frank Lloyd Wright on the power of atmosphere: "Whether people are fully conscious of this or not, they actually derive *countenance* and *sustenance* from the 'atmosphere' of things they live in and with."[25] This view of the dominance of unified entities over "elements" casts serious doubt on the prevailing elementarist theories and teaching methods in education.

THE ATMOSPHERIC SENSE

I have become so impressed with the power of our atmospheric judgment that I want to suggest that this capacity could be named our sixth sense. Thinking only of the five Aristotelian senses in architecture fails to acknowledge the true complexity of the systems through which we are connected to the world. Steinerian philosophy, for instance, deals with twelve senses,[26] whereas a recent book, *The Sixth Sense Reader*, identifies more than thirty categories of sensing through which we relate to and communicate with the world.[27] This idea of a wider human sensorium underlines the fact that our being-in-the-world is much more complex and refined than we tend to understand. That is why understanding architecture solely as a visual art form is hopelessly reductive. Besides, instead of thinking of the senses as isolated systems, we should become more interested in and knowledgeable about their essential interactions and crossovers. Merleau-Ponty emphasizes this essential unity and interaction of the senses: "My perception is ... not a sum of visual, tactile, and auditive givens: I perceive in a total way with my whole being. I grasp a unique structure of the thing, a unique way of being, which speaks to all my senses at once."[28] This flexibility and dynamic of our interaction with the world is one of the important things that neuroscience can illuminate for us. The craft of architecture is deeply embedded in this human sensory and mental complexity.

This criticism of the reductive isolation of the senses also applies to the common understanding of intelligence as a singular intellectual capacity. Contrary to the common understanding of intelligence as a definite cerebral category, psychologist Howard Gardner suggests seven categories of intelligence, namely linguistic, logical-mathematical, musical, bodily-kinesthetic, spatial, interpersonal, and intrapersonal intelligences, to which he later adds three further categories: naturalistic, ethical, and spiritual intelligences.[29] I would add four further categories to Gardner's list: emotional, aesthetic, existential, and atmospheric intelligences. So, we may well have a full spectrum of a dozen modes of intelligence instead of the single quality targeted by IQ tests. The complex field of intelligence also suggests that architectural education, or education at large, faces a much wider task, and at the same time possesses far greater potential, than standard pedagogy has thus far accepted. Education in any creative field must start primarily with the student's sense of self, as only a firm sense of identity and self-awareness can serve as the core around which observation, knowledge, and eventually wisdom can evolve and condense.

HUMAN BIOLOGICAL HISTORICITY

We also need to accept the essential historical and embodied essence of human existence, experience, cognition, and memory. In our bodies we can still identify the remains of the tail from our arboreal life; the pink triangular area in our eye corners, the *plica semilunaris*, is the remnant of our horizontally moving eyelid from the saurian age; and even the traces of gills derive from our aquatic life hundreds of millions of years ago. We certainly have similar imprints in our mental constitution that derive from our biological and cultural historicity; one aspect of such deeply concealed memory was pointed out by Sigmund Freud and Carl G. Jung—namely, the archetype.[30] I want to add here that Jung defined archetypes dynamically, as certain tendencies for distinct images to evoke certain types of associations and feelings. So, even archetypes are not concrete or given "building blocks" in artistic creation—as postmodernists seemed to believe—but dynamic and interacting mental forces with lives of their own.

Architecture, also, has its roots and mental resonances in our biological historicity. Why do we all sense profound pleasure when sitting by an open fire, if not because fire has offered our predecessors safety, pleasure, and a heightened sense of togetherness for some fifty thousand years? Vitruvius, in fact, dates the beginning of architecture to the domestication of fire. The taming of fire actually gave rise to unexpected changes in the human species and its behavior. "Control over fire changed human anatomy and physiology and became encoded in our evolving genome," argues Stephen Pyne, who attributes the changes in human teeth and intestinal structures to the consequences of eating cooked food.[31] Some linguistic scholars have suggested that language also originates in the primordial act of gathering around the fire. Such biopsychological heritage, especially the polarity of "refuge" and "prospect," has been observed in Frank Lloyd Wright's houses by Grant Hildebrandt.[32] The writer suggests that the master architect intuited the meaning of this spatial polarity decades before ecological psychology touched upon the phenomenon. The studies of the American anthropologist Edward T. Hall, in the 1960s, revealed unbelievably precise unconscious mechanisms in the use of space and its culture-specific parameters.[33] "Proxemics," the new field of study Hall initiated, is based on such unconscious spatial mechanisms. He acknowledges the external communication between our endocrine glands, in opposition to the prevailing scientific view that these glands

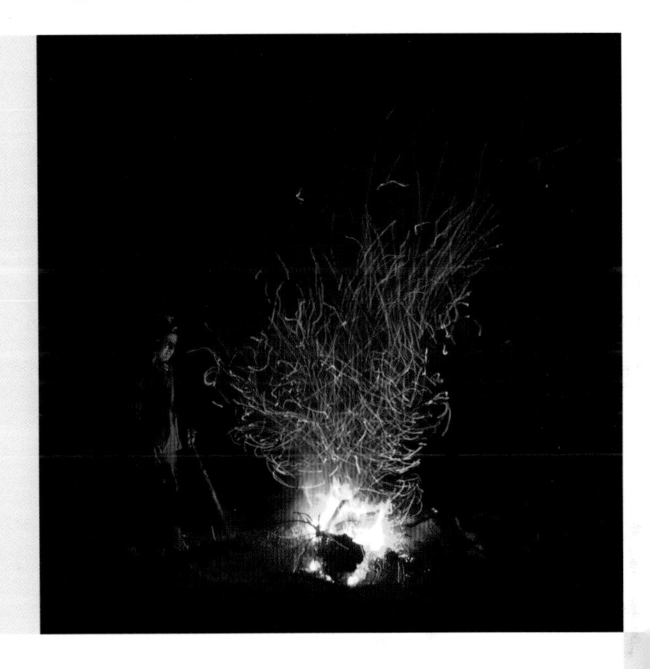

3.8 The domestication of fire strengthened the social bond and permitted interaction between the members of the group during the period of darkness. Vitruvius dates the origins of architecture to the taming of fire, while some contemporary scholars suggest that the unifying impact of fire served as a factor in the evolution of language.

have only internal metabolic functions—yet another example of the ambiguity of the boundary of the self.[34] In her chapter "Nested Bodies" in this book, Sarah Robinson points out the bioelectric and magnetic fields originating in the body, which extend our bodies beyond the boundary of the skin.[35] Finally, philosopher Martin Jay's remark "With vision we touch the sun and the stars"[36] turns us into true cosmological beings.

Such studies are surely only beginning to reconnect modern man, *Homo faber*, back to his biological roots; and we look to neuroscience to valorize the internal workings of these physiological, genetic, and instinctual capacities and reactions. Neurological studies can reveal the neural ground for our fundamental spatial and environmental pleasures and displeasures—as well as our feelings of comfort, safety, and fear.

Merleau-Ponty makes the significant remark: "The painter takes his body with him. ... Indeed, we cannot imagine how a mind could paint."[37] The same must certainly be said about architects, as our craft is unavoidably constituted in our embodied manner of existence; and architecture articulates that very mode of being. In my way of thinking, architecture is more an art of the body and existential sense than one of the eye (even vision

3.9 Alvar Aalto's "extended Rationalism" and fusion of opposites; the living space of the Villa Mairea merges images of tectonic architectural space and amorphous forest space, modern utopia and peasant tradition. Alvar Aalto, Villa Mairea, Noormarkku, Finland, 1938–1939.

serves our existential sense of being)—more one of emotive and unconscious feelings than rational deduction. This is where the logocentric and overintellectualized theorizing of architecture, so popular in the recent past, has gone decisively wrong. But, again, neuroscience can probably valorize these interactions, hierarchies, and priorities. I believe that research in the biological and neurosciences will confirm that our experiences of architecture are in fact grounded in the deep and unconscious layers of our mental life.

I am not speaking against attempts to grasp the structure or logic of experiential phenomena; I am merely concerned about a reductivist or biased understanding of architectural phenomena. The study of artistic phenomena also calls for appropriate methods of study. In the mid-1930s, Alvar Aalto wrote about "an extended Rationalism," and urged architects to expand rational methods even to the psychological (Aalto used the terms "neurophysiology" and "psychophysical field") and mental areas.[38] Both Wright's and Aalto's masterworks are examples of an architecture that benevolently embraces us, which can hardly be explained intellectually. This is an architecture that is directly connected with our human nature by the architect's own intuited wisdom. No doubt, great architects throughout history have always intuitively grasped the essence of human mental life—both individual and collective. Does not the notion of genius imply capacities of intuiting entities, interrelations, and causalities beyond the boundaries of established knowledge?

We are mentally and emotionally affected by works of architecture and art before we understand them; or, indeed, we usually do not "understand" them at all. I would argue that the greater the artistic work, the less we understand it intellectually. Do we really understand Michelangelo's Rondadini *Pietà*, Giorgione's *Tempest*, or Rembrandt's portraits? No—they will always remain unexplainable jewels of our experiential world. A distinct mental short-circuiting between a lived emotional encounter and intellectual

3.10 Tintoretto's painting of a dramatic subject projects a strong atmosphere that unifies the multitude of narrative and pictorial ingredients into a cohesive and emotionally embracing ensemble. The parts cannot be distinguished from the impact of the whole. Jacopo Tintoretto (Jacopo Comin), *Crucifixion*, 1565. 518 × 1224 cm. Scuola di San Rocco, Venice.

understanding is a constitutive characteristic of the artistic image. Jean-Paul Sartre points out the essential fusion of the object and its experience in the artistic encounter: "Tintoretto did not choose that yellow rift in the sky above Golgotha to signify anguish or to provoke it. Not sky of anguish or anguished sky; it is an anguish become thing, an anguish which has turned into yellow rift of sky. ... It is no longer readable."[39] In fact, art is not about understanding at all; an artistic image is an existential encounter which momentarily reorients our entire sense of being: just think of the mysterious powers of music. Great works possess a timeless freshness; they project their enigmas always anew—making us feel each time that we are experiencing the work for the first time. I like to revisit architectural and artistic masterpieces around the world to repeatedly encounter their magical sense of newness and freshness. The greater the work, the stronger its resistance to time. As Paul Valéry suggests, "An artist is worth a thousand centuries."[40] The hypnotic power of the cave paintings testifies to this longevity of artistic images. The interaction of newness and the primordial in the human mind is yet another aspect of the artistic and architectural image that can be understood through neuroscientific research, I believe. Our neural system seems to be activated by newness, and we seek novel stimuli, whereas the deepest emotive impact arises from the primal layers of our neural system and memory. We humans are essentially creatures suspended between the past and the future more poignantly than other forms of life—it is the task of art to mediate between these polarities.

ARTISTS AS "NEUROLOGISTS"

Semir Zeki, a neurologist who studies the neural ground of artistic image and effect, considers a high degree of ambiguity—such as the unfinished imagery of Michelangelo's slaves, or the ambivalent human narratives of Johannes Vermeer's paintings—to be essential to the greatness of these works.[41] In reference to the great capacity of profound artists to evoke, manipulate, and direct emotions, he posits the surprising argument: "Most painters are also neurologists ... they are those who have experimented with and, without ever realizing it, understood something about the organization of the visual brain, though with the techniques that are unique to them."[42] This statement interestingly echoes an argument of the Dutch phenomenologist-therapist J. H. Van den Berg: "All painters and poets are born phenomenologists."[43] Artists and architects are phenomenologists in the sense of being capable of "pure looking," an unbiased and "naive" manner of encountering things. In fact, Bachelard advises practitioners of the phenomenological approach "to be systematically modest" and "to go in the direction of maximum simplicity."[44] A recent book, *Proust Was a Neuroscientist* by Jonah Lehrer, popularizes this topic, arguing that certain masterly artists, such as Walt Whitman, Marcel Proust, Paul Cézanne, Igor Stravinsky, and Gertrude Stein, anticipated some of today's crucial neurological findings through their art more than a century ago.[45] In his important books *The Architect's Brain* and *Architecture and Embodiment*, Harry F. Mallgrave connects the latest findings in the neurosciences with the field of architecture directly in accordance with the objective of this book.[46]

In *Inner Vision*, Semir Zeki suggests the possibility of "a theory of aesthetics that is biologically based."[47] Having studied animal building behavior and the emergence of "aesthetically" motivated choices in the animal world for forty years, I have no doubt about this. What else could beauty be than nature's powerful instrument of selection in the process of evolution? Joseph Brodsky assures us of this with the conviction of a poet: “The purpose of evolution, believe it or not, is beauty.”[48]

It is beyond doubt that nature can teach us great lessons about design, particularly about ecologically adapted design and dynamic processes. This can be seen in emerging fields of study, such as bionics and biomimicry. Several years ago, I had the opportunity to participate in a conference in Venice entitled “What Can We Learn from Swarming Insects?” organized by the European Center for Living Technologies. The participants were biologists, mathematicians, computer scientists, and a couple of architects. The purpose of the

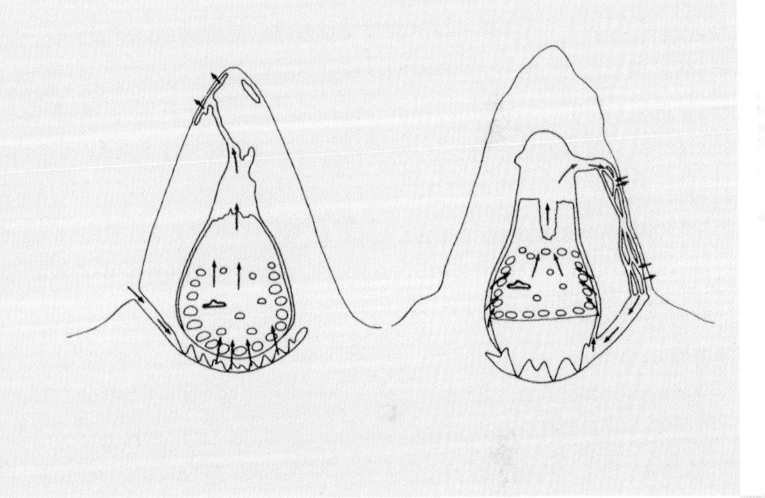

3.11 Miracles of functional design in the animal world: *Microtermes bellicosus* termite nest from the Ivory Coast (left) and Uganda (right). The arrows indicate the directions of air flows. Termites of a single species living in the coastal and inland climatic conditions seem to be able to construct two different air-conditioning systems for their nest depending on the climate.

encounter was to gain understanding, through recent research findings and computer simulations, of the miraculous capacities of ants, termites, bees, and wasps to construct perfectly adapted nests and wider environmental systems, such as fungus farms and covered road networks. So far, the chain of collective and instinctual actions that enable termites to construct a vault has been simulated, but the embodied collective knowledge that enables them to construct their nest as an artificial lung to sustain the life of a community of millions of individuals remains far beyond our understanding.[49] We can surely expect more of such deliberations in the future. Edward O. Wilson, the world’s leading myrmecologist and pioneer of biophilia, “the new ethics and science of life,” makes the dizzying argument that “the superorganism of a leaf-cutter ant nest is a more complex system in its performance than any human invention, and unimaginably old.”[50]

In his study on the neurological ground of art, Zeki argues that "art is an extension of the functions of the visual brain in its search for essentials."[51] I see no reason to limit this idea of extension, or externalization, only to the visual field. I believe that art provides momentary extensions of the functions of our perceptual systems, consciousness, memory, emotions, and existential "understanding." The great gift of art is to permit us ordinary mortals to experience something through the perceptual and emotive sensibility of some of the greatest individuals in human history. We can feel through the neural subtlety of Michelangelo, Bach, and Rilke, for instance. And again, we can undoubtedly make the same assumption about meaningful architecture; we can sense our own existence amplified and sensitized by the works of great architects from Ictinus and Callicrates to Frank Lloyd Wright and Louis Kahn.

The role of architecture as a functional and mental extension of our capacities is clear, and in fact Richard Dawkins has described various aspects of this notion among animals in his book *The Extended Phenotype*;[52] he suggests that such fabricated extensions of biological species should be made part of the phenotype of the species in question. So, dams and water regulation systems should be part of the phenotype of the beaver, and the astounding nets of the spider. Works of meaningful architecture intuitively grasp the essence of human nature and behavior, in addition to being sensitive to the hidden biological and mental characteristics of space, form, and materiality. By intuiting this knowledge, sensitive architects are able to create places and atmospheres that make us feel safe, comfortable, invigorated, and dignified without being able to conceptually theorize their skills at all. In this context, I have earlier used the notion "a natural philosophy of architecture," a wisdom that arises directly from an intuitive and lived understanding of human nature, and architecture as an extension of that very nature. Simply, great architecture emanates unspoken but contagious existential wisdom.

THE GIFT OF THE IMAGINATION

The imagination is arguably the most human of our capacities. Although it is often considered to be a kind of daydreaming, and sometimes even as something suspect, our most basic activities, such as perceiving and memorizing places, situations and events, rely on our imagination. The acts of experiencing and memorizing are embodied acts, which evoke imaginative realities with specific meanings. The existence of our ethical sensibility alone calls for imaginative skills, as we could not evaluate our alternative behavioral choices without the ability to imagine their consequences. Recent studies have revealed that the acts of perceiving and imagining take place in the same areas of the brain;

consequently, these act are closely related.[53] "Every act of perception is an act of creation," argues neurophysiologist Gerald Edelman, as Sarah Robinson notes.[54] Or, "attention is a deeply creative act," as McGilchrist points out.[55] Perceptions call for imagination, as percepts are not automatic products of the sensory mechanism; they are essentially interpretations, projections, creations, and products of intentionality and imagination. We could not even see light without our "inner light" and "formative visual imagination," argues the physicist Arthur Zajonc.[56] To conclude: "Reality is a product of the most august imagination," as the poet Wallace Stevens suggests.[57]

We do not judge environments merely by our senses, we also test and evaluate them through our imagination. Comforting and inviting settings inspire our unconscious imagery, daydreams, and fantasies. Sensuous settings sensitize and eroticize our relationship with the world. As Bachelard argues, the "chief benefit of the house [is that] the house shelters daydreaming, the house protects the dreamer, the house allows one to dream in peace. ... The house is one of the greatest powers of integration for the thoughts, memories and dreams of mankind."[58]

BODY AND IMAGINATION IN THE ARTS

I have found the study of other art forms very illuminating for the understanding of the mental phenomena in architecture, because the subtleties of our mental reactions are usually confused or suppressed by more practical and rational considerations in the craft of architecture. The processes of literary imagination are interestingly discussed in Elaine Scarry's recent book *Dreaming by the Book*. In her view, great writers—from Homer, Flaubert, and Rilke to today's masters of literature, such as Seamus Heaney—have intuited, through words, how the brain perceives images. She explains the vividness of a profound literary text: "In order to achieve the 'vivacity' of the material world, the verbal arts must somehow also imitate its 'persistence' and, most crucially, its quality of 'givenness.' It seems almost certainly the case that it is the 'instructional' character of the verbal arts that fulfills this mimetic requirement for 'givenness.'"[59] It is the experience of givenness, naturalness, and inevitability that is missing in today's architecture of intellectual games and formal invention.

The Czech writer Bohumil Hrabal vividly describes the concreteness and embodied nature of literary imagination: "When I read, I don't really read: I pop up a beautiful sentence in my mouth and suck it like liqueur until the thought dissolves in me like alcohol, infusing my brain and heart and coursing on through the veins to the root of each blood vessel."[60] Permit me to give yet another example of the embodied nature of poetry. Charles Tomlinson, a poet, observes the bodily basis even of the practices of painting and poetry: "Painting wakes up the hand, draws in your sense of muscular coordination, your sense of the body, if you like. Poetry, also, as it pivots on its stresses, as it rides forward over the line-endings, or comes to rest at pauses in the line, poetry also brings the whole man into play and his bodily sense of himself."[61] Surprisingly, Henry David Thoreau already grasped the significance of the body in poetry: "The poet creates the history of his own body."[62] It is, of course, clear that architecture is the art form that "brings the whole man into play and his bodily sense of himself," exactly in accordance with

3.12 Every significant work of art is a complete microcosm, a metaphoric universe of its own. Morandi's still lifes of timid objects on a table top turn into instruments for intense metaphysical contemplation. Giorgio Morandi, *Still Life*, 1958. Oil on canvas, 25 × 40 cm. Private collection, Bologna.

Tomlinson's description above. Architecture is born of the body, and when we experience profound architecture we return to the body.

As our age seems to value fictions, fantasies, and virtual realities, I wish to include an example of the role of the sense of reality in artistic works. Jorge Luis Borges gives us important advice concerning the requirement for a sense of reality and artistic plausibility: "Reality is not always probable, or likely. But if you're writing a story, you have to make it as plausible as you can, because otherwise the reader's imagination will reject it."[63] Regardless of today's obsession with the fantastic image, architecture is similarly an art form of reality, not fantasy; architecture's task is to reinforce our sense of the real and, through doing that, to liberate our senses and imagination.

Profound works of architecture are not merely imaginary and aestheticized settings or objects; they are complete microcosmic worlds. "If a painter presents us with a field or a vase of flowers, his paintings are windows, which are open on the whole world," Jean-Paul Sartre avers.[64] A Giorgio Morandi painting with a couple of shy vases and glasses on a table is in fact a metaphysical deliberation which invites the viewer to zoom into the

3.13 Louis I. Kahn, Library and Dining Hall, Phillips Exeter Academy, Exeter, New Hampshire, 1965–1972. All imposing works of architecture are spatial mandalas and metaphoric representations of the world. They enable us to feel "how the world touches us," as Merleau-Ponty said of the paintings of Paul Cézanne.

most haunting question of all, that of being: why are there objects and things rather than not? Architecture, also, mediates similarly deep narratives of culture, place, and time, and it is essentially an epic art form, expressive of human life and culture. The content and meaning of art—even the most condensed poem, minimal painting, or simplest hut—is epic in the sense of being a lived metaphor of human existence in the world.

I wish to end with one of the most impressive statements about the mental quality of art that I have read. This poetic requirement distills my arguments about essential artistic condensation, and it also applies fully to architecture. As the master sculptor Constantin Brâncuşi advises us: "The work must give immediately, at once, the shock of life, the sensation of breathing."[65]

NOTES

1. Richard Rorty, *Philosophy and the Mirror of Nature* (Princeton: Princeton University Press, 1979), 239.

2. Jean-Paul Sartre, *The Emotions: An Outline of a Theory* (New York: Carol Publishing, 1993), 9.

3. Frank Lloyd Wright quote on a teacup purchased at the Taliesin West book and gift shop.

4. Karsten Harries, "Building and the Terror of Time," *Perspecta: The Yale Architecture Journal*, no. 19 (1982).

5. Gaston Bachelard, *The Poetics of Space* (Boston: Beacon Press, 1969), 46.

6. Ibid., 7.

7. Rudolf Wittkower, *Architectural Principles in the Age of Humanism* (New York: Random House, 1965), 16.

8. Keijo Petäjä in numerous conversations with the author during the 1970s. The Finnish original reads: "Arkkitehtuuri on rakennettua mielen tilaa."

9. Maurice Merleau-Ponty, "The Intertwining—The Chiasm," in Merleau-Ponty, *The Visible and the Invisible*, ed. Claude Lefort (Evanston: Northwestern University Press, 1969).

10. Maurice Merleau-Ponty describes the notion of the flesh: "My body is made of the same flesh as the world ... and moreover ... this flesh of my body is shared by the world" (ibid., 248); and "The flesh [of the world or my own] is ... a texture that returns to itself and conforms to itself" (146).

11. Maurice Merleau-Ponty, *Signs* (Evanston: Northwestern University Press, 1982), 56.

12. Salman Rushdie, "Eikö mikään ole pyhää? [Is nothing sacred?]," *Parnasso* (Helsinki) 1 (1996): 8. Trans. Juhani Pallasmaa.

13. Balthus, in Claude Roy, *Balthus* (New York: Little, Brown, 1996), 18.

14. Balthus, in *Balthus in His Own Words: A Conversation with Cristina Carrillo de Albornos* (New York: Assouline, 2001), 6.

15. Maurice Merleau-Ponty, as quoted in Iain McGilchrist, *The Master and His Emissary: The Divided Brain and the Making of the Western World* (New Haven: Yale University Press, 2010), 409.

16. Alvar Aalto, "Taide ja tekniikka" (Art and Technology), lecture, Academy of Finland, October 3, 1955, in Göran Schildt, *Alvar Aalto Luonnoksia* (Helsinki: Otava Publishers, 1972), 87–88. Trans. Juhani Pallasmaa.

17. Sir Colin St. John Wilson, "Architecture—Public Good and Private Necessity," *RIBA Journal* (March 1979): 107–115.

18. Gaston Bachelard, *The Philosophy of No: A Philosophy of the New Scientific Mind* (New York: Orion Press, 1968), 16.

19. Ibid.

20. Joseph Brodsky, "An Immodest Proposal," in Brodsky, *On Grief and Reason* (New York: Farrar, Straus and Giroux, 1997), 206.

21. Paul Valéry, "Eupalinos, or the Architect," in Valéry, *Dialogues* (New York: Pantheon Books, 1956), 83; original emphasis.

22. Iain McGilchrist, "Tending to the World," chapter 5 below.

23. John Dewey, *Art as Experience* (1934; New York: Perigee Books, 1980), 35; original emphasis.

24. Ibid., 73.

25. Frank Lloyd Wright; Sarah Robinson's information in a letter to the author, 20 January 2012. Original emphasis.

26. Albert Soesman, *Our Twelve Senses: Wellsprings of the Soul* (Worcester: Hawthorn Press , Stroud, Gloucestershire, UK, 1998).

27. *The Sixth Sense Reader*, ed. David Howes (New York: Berg, 2011), 23–24.

28. Maurice Merleau-Ponty, "The Film and the New Psychology," in Merleau-Ponty, *Sense and Non-sense* (Evanston: Northwestern University Press, 1964), 50.

29. Howard Gardner, *Intelligence Reframed: Multiple Intelligences for the 21st Century* (New York: Basic Books, 1999).

30. See Carl G. Jung et al., eds., *Man and His Symbols* (New York: Doubleday, 1968), 57.

31. Stephen J. Pyne, *Fire* (London: Reaktion Books, 2012), 47.

32. Grant Hildebrandt, *The Wright Space: Pattern and Meaning in Frank Lloyd Wright's Houses* (Seattle: University of Washington Press, 1992).

33. Edward T. Hall, *The Silent Language* (New York: Anchor Press, 1959); and *The Hidden Dimension* (New York: Doubleday, 1966).

34. Hall, *The Hidden Dimension*, 33–34. The writer refers to endocrinological research by Parkes and Bruce.

35. Sarah Robinson, "Nested Bodies," chapter 7 below.

36. Martin Jay, as quoted in David Michael Levin, ed., *Modernity and the Hegemony of Vision* (Berkeley: University of California Press, 1993), 14.

37. Maurice Merleau-Ponty, *The Primacy of Perception* (Evanston: Northwestern University Press, 1964), 162.

38. Alvar Aalto, "The Humanizing of Architecture," *Technology Review* (November 1940), as reprinted in *Alvar Aalto: Sketches*, ed. Göran Schildt (Cambridge, MA: MIT Press, 1979), 77, 78.

39. Jean-Paul Sartre, *What Is Literature?* (Gloucester, MA: Peter Smith, 1978), 3.

40. Valéry, *Dialogues*, xiii.

41. Semir Zeki, *Inner Vision: An Exploration of Art and the Brain* (Oxford: Oxford University Press, 1999), 22–36.

42. Ibid., 2.

43. J. H. Van den Berg, as quoted in Bachelard, *The Poetics of Space*, xxiv.

44. Bachelard, *The Poetics of Space*, xxi, 107.

45. Jonah Lehrer, *Proust Was a Neuroscientist* (New York: Houghton Mifflin, 2008).

46. Harry Francis Mallgrave, *The Architect's Brain: Neuroscience, Creativity, and Architecture* (Chichester: Wiley-Blackwell, 2010); and *Architecture and Embodiment: The Implications of the New Sciences and Humanities for Design* (Abingdon, UK: Routledge, 2013).

47. Zeki, *Inner Vision*, 1.

48. Brodsky, "An Immodest Proposal," 208.

49. See Juhani Pallasmaa, *Eläinten arkkitehtuuri—Animal Architecture* (Helsinki: Museum of Finnish Architecture, 1995).

50. Edward O. Wilson, *Biophilia: The Human Bond with Other Species* (Cambridge, MA: Harvard University Press, 1984), 37.

51. Zeki, *Inner Vision*, 1.

52. Richard Dawkins, *The Extended Phenotype* (Oxford: Oxford University Press, 1982).

53. Ilpo Kojo, "Mielikuvat ovat aivoille todellisia [Images are real for the brain]," *Helsingin Sanomat*, Helsinki, March 26, 1996. The article refers to research at Harvard University under the supervision of Dr. Stephen Kosslyn in the mid- 1990s.

54. Gerald Edelman, "From Brain Dynamics to Consciousness: How Matter Becomes Imagination," Marschak Memorial Lecture at UCLA, February 18, 2005; as quoted in Sarah Robinson's chapter in this volume.

55. McGilchrist, "Tending to the World," in this volume.

56. Arthur Zajonc, *Catching the Light: The Entwined History of Light and Mind* (Oxford: Oxford University Press, 1995), 5.

57. Quoted in Lehrer, *Proust Was a Neuroscientist*, vi.

58. Bachelard, *The Poetics of Space*, 6.

59. Elaine Scarry, *Dreaming by the Book* (Princeton: Princeton University Press, 2001), 30.

60. Bohumil Hrabal, *Too Loud a Solitude* (New York: Harcourt, 1990), 1.

61. Charles Tomlinson, "The Poet as Painter," in J. D. McClatchy, ed., *Poets on Painters* (Berkeley: University of California Press, 1990), 280.

62. Henry David Thoreau, as quoted in Lehrer, *Proust Was a Neuroscientist*, 1.

63. Jorge Luis Borges, *Borges on Writing*, ed. Norman Thomas di Giovanni, Daniel Halper, and Frank MacShane (Hopewell: Echo Press, 1994), 45.

64. Sartre, *What Is Literature?*, 272.

65. Constantin Brâncuşi as quoted in Eric Shanes, *Brancusi* (New York: Abbeville Press, 1989), 67.

4

TOWARD A NEUROSCIENCE OF THE DESIGN PROCESS

Michael Arbib

While I have come up with three different ways in which neuroscience might inform the work of architects, here I will emphasize the *neuroscience of the design process*, whose central question is: "What can we understand about the brain of the architect as he or she designs a building?" I will offer only a preliminary analysis, but hope to encourage further thinking about how the design process can be illuminated more and more by future research in neuroscience. In addition, I will briefly introduce the two other areas: one is the *neuroscience of the experience of architecture*: not what goes on in the architect, but what goes on in the person experiencing a building; the other is what I call *neuromorphic architecture*.[1] But let me first briefly offer a perspective on neuroscience.

A FEW WORDS ABOUT NEUROSCIENCE

Much of the discussion of neuroscience starts with a quality such as an action, vision, memory, or empathy of which we are aware without knowledge of neuroscience, and then shares the excitement of experiments that *correlate* degrees of people's involvement in such a quality with degrees of activation in brain regions as seen during brain imaging. I want to probe more deeply to ask how circuitry within the brain mediates our action, perception, and memory—all in relation to our interactions with and experiences of both the physical and social worlds in which we are immersed. In the context of this chapter I can only point you in the general direction of such studies—a reasonably full yet accessible exposition may be found in my book *How the Brain Got Language.*[2] My concern is with "how the brain works."

In 1905, the neuroanatomist Korbinian Brodmann stained regions of the brain and then looked through a light microscope to see how the layered structure of cerebral cortex differed from place to place. He then assigned numbers to cortical regions which had a distinctive pattern of layering. This procedure correlated different regions in the brain with their specific functions. It is important to remember, however, that the brain is not a set of different boxes each doing a separate job—rather, multiple regions compete and cooperate, making their own specialized contributions to a range of cognitive functions.

Other neuroanatomists, starting with Ramón y Cajal in Spain (who also trained as an artist and was a pioneer in color photography), moved from staining that showed patterns of layering to staining a sampling of individual neurons in exquisite detail. Figures 4.1 and 4.2 show several of Cajal's drawings, which are artistic and evocative as well as displaying great insight into the way neurons work together. Figure 4.3, by the Hungarian neuroanatomist János Szentágothai, shows a selection of neural circuits of the cerebral cortex revealed in their beautiful particularity. This is the level at which I worry most about neuroscience: I am interested in what is going on, not at the level of the gross areas delineated by Brodmann, but in terms of the circuitry, dramatically different in different brain regions, whose patterns of activation mediate our perception, our actions, our memories, our desires.

But of course there are many other levels as well. Below the neuron we have *synapses*—the connection points between neurons. These are elaborate molecular and membrane structures. For our purposes here, it will suffice to simply think of synapses as *loci* of change. Molecules and membranes may be too detailed for the time being; our main concern is to understand how experience shapes the way one neuron talks to other

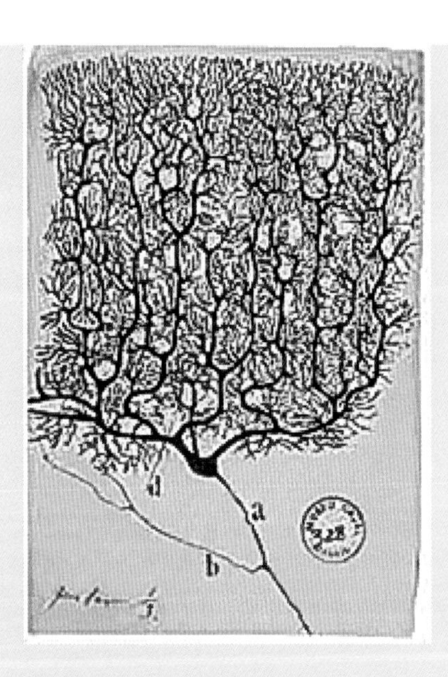

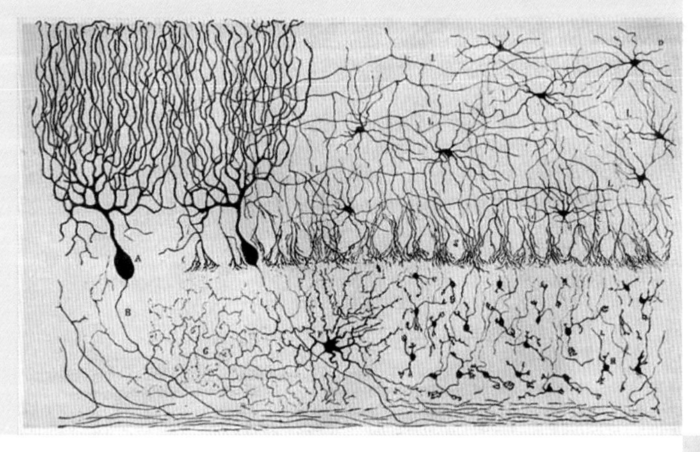

4.1 AND 4.2 Some examples of the scientific artistry of Santiago Ramón y Cajal. At left, a Purkinje cell, the only type of cell whereby the cerebellar cortex communicates with the rest of the brain; at right, various neurons of the cat brain.

neurons by adapting the strength of the synapses between them. I have already discussed *circuits*, which are above neurons. And above circuits are the *schemas*, which bridge our psychology, our experience and behavior, with what neural circuits are doing. The notion of "action-oriented perception"[3] is crucial here: one should look at the perceptual systems of the brain not as ends in themselves (reconstructing the world in our heads) but rather in terms of our ongoing courses of action, a point emphasized in the notion of the *action-perception cycle*.[4] What we do depends on what we have perceived, but what we perceive depends on what we do—and our actions include exploration in search of knowledge of the world relevant to our unfolding goals and plans.

Moving up beyond the study of an individual brain, we need to assess how we, as *persons*, experience not only the physical but also the social world. What is it that makes us a person rather than a bunch of brain areas with a bunch of muscles? Above the level of the individual, our membership in a social group comes into play: social interactions, as well as embodied interactions, shape who we are.

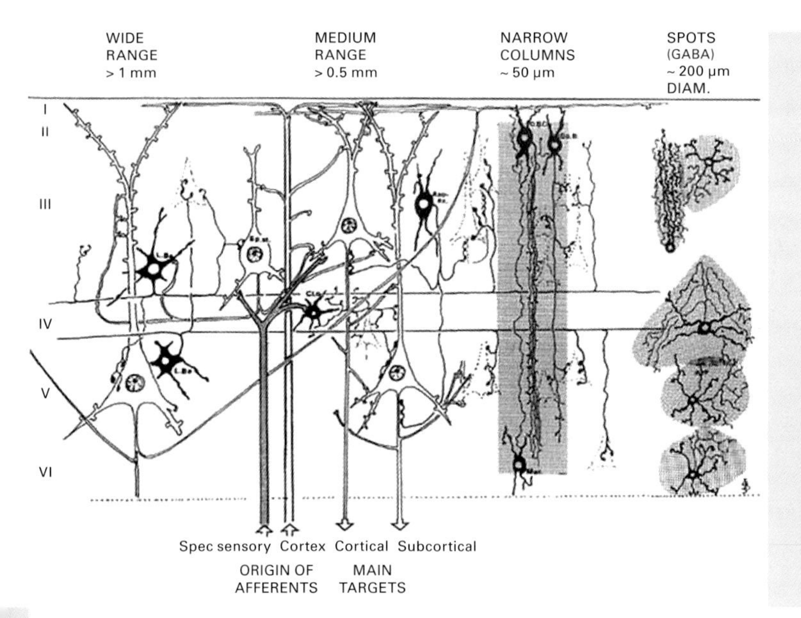

4.3 A schematization of the neural circuitry of cerebral cortex, by János Szentágothai. In the real brain, many more cells are packed into a patch corresponding to this figure, and the distribution of cells differs markedly from region to region—with these differences providing the basis for the distinctions drawn by Brodmann.

SCHEMAS

A particular challenge for any neuroscience for architecture is to bridge from the language of psychology to that of neural circuitry. The way I have approached this (and not just for architecture) is to develop a neurally inspired notion of *schemas*, inspired in part by the genetic epistemology of Jean Piaget,[5] but influenced more strongly by an attempt to advance *neuroethology*, the study of the brain mechanisms underlying animal behavior.[6] I distinguish perceptual schemas from motor schemas, while acknowledging that many bridging schemas are required, especially in the human ability to contemplate concepts of increasing abstraction.

A *perceptual schema* is a process for recognizing an apple or a person's face, a chair or a wall. But it is not enough to recognize one object at a time. To make sense of your environment you have to be able to recognize many different objects and their spatial relationships. On the other hand, *motor schemas* provide the ability to carry out the actions that have been determined through the continued operation of the action-perception cycle. Note the plural—we seldom execute a single action, but rather a "coordinated control program" that coordinates the various motor schemas, modulating their activity as perceptual schemas update their representation of the current state of the actor's interaction with the environment.

This notion was inspired by figure 4.4 from Marc Jeannerod and Jean Biguer, which shows that as you reach to grasp, you are not only getting your hand in the right place but also preshaping your hand in preparation for grasping the object. The idea of the coordinated control programs is that a perceptual schema must recognize object location to direct the motor schema for grasping; another perceptual schema must recognize object size and orientation to direct the motor schema for grasping; and all four schemas must be coordinated for a successful reach-to-grasp.

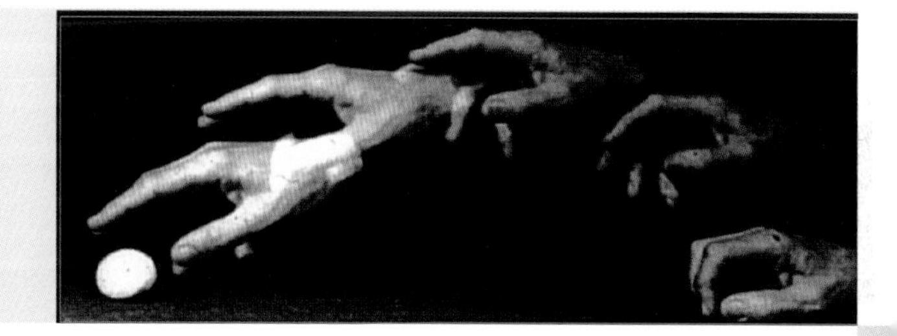

4.4 Preshaping while reaching to grasp (Marc Jeannerod and Jean Biguer, 1982).

For us, phenomenologically, the reach-to-grasp is a seamless whole combining the location and shape of the object into a single integrated action of the hand. Yet if people have damage to a certain part of the cerebrum, the action is no longer seamless. They are unable to use visual information to preshape—instead they have to open their hand maximally and hold its extended shape as they move to the object, and then use the sense of touch to shape their hand to grasp it. A patient with a different lesion may preshape her hand appropriately to the size of an object as she reaches to grasp it—yet be unable to tell you (either verbally or through pantomime) how big the object is. This introduces a crucial theme. *What may seem like a seamless unity to our introspection may in fact involve different processes in the brain being coordinated and orchestrated.* The notion, then, is that as we study the neuroscience of processes relevant to the experience and design of buildings, we may learn new ways of dissecting those processes and build upon the resultant insights to provide new tools (whether conceptual or computational) for design.

We now turn from the effects of brain damage on human action and perception to detailed studies using fine microelectrodes to measure the activity of single neurons in the monkey brain.[7] Two parts of the macaque brain, AIP and F5, are of special importance, but for our present purposes the meaning of these terms is not essential. In AIP, Sakata and his colleagues in Tokyo found cells whose firing did not depend on what type of object the monkey was looking at but only its *affordances*, its possibilities for being grasped. Similarly, when Rizzolatti and his colleagues in Parma recorded from area F5, they found cells whose activity correlated with the *motor schema* of the monkey's grasp—a precision grip might be correlated with one set of cells, power grasps with others, and so on.

This term, affordances, is an important concept in Sarah Robinson's book *Nesting*:[8] we get this word from J. J. Gibson, who used it to identify the opportunities for action that our brains register without necessarily being consciously aware of the object which affords those actions. The easiest way to understand this is to imagine that you are walking down a street when your peripheral vision picks up a possible collision and you move aside to avoid it rather than turning your attention to identify the obstacle—and then later you get an aggrieved phone call from a friend, who was walking down the street at the same moment, asking why you ignored them. To which you reply, "What?" You have one part of your visual system extracting affordances for immediate action, while the more central system answers the question "Who is that, what is that, how does that enter into my decision-making?" We can understand this level of human behavior all the more fully because we can tease apart the neural details in the monkey. Much architecture is theorized and taught solely in terms of focused vision, but we are embraced by space by means of our peripheral perception.

Turning to the perception of visual scenes, recall René Magritte's 1959 painting *Le Château des Pyrénées*. If you look only at its top half, you get an immediate impression of a mountaintop with a castle on it. But when you see the whole picture, you are confounded—what had appeared to be a mountain is revealed to be an immense boulder suspended above (or crashing into?) the waves below. What is interesting here is that in general we live in a consistent world, so that our schemas (interacting states of activity in our brains) can "nudge" each other, changing the shapes they represent, redoing their parametric descriptions, so that between them they provide a coherent interpretation of the scene. As a surrealist, Magritte painted each part of the painting totally realistically. There is no way you can fail to see that there is an ocean at the bottom of his painting, just as there is no way you can fail to see the bottom of the boulder above it; yet from

4.5 René Magritte, *Le Château des Pyrénées*, 1959.

the top, the view of the castle confounds the notion that it sits atop a boulder, rather than a mountainside. Magritte creates a form of cognitive dissonance. I use this example to illustrate that in most cases, moving your eyes around the scene builds up an integrated view. In general this will be a consistent view of what is out there, unless you are faced with Magritte's surrealism, whose intent was to deliberately confuse. Different people, of course, have different experiences, and these experiences shape different perceptual schemas. These differences are also partially due to the fact that during the action-perception cycle, diverse motor schemas are coupled with and underpinned by the individual's skill set.

NEUROSCIENCE OF THE EXPERIENCE OF ARCHITECTURE

One of the three areas into which I divide the conversation between neuroscience and architecture is the *neuroscience of the experience of architecture*. The Academy of Neuroscience for Architecture was founded by John Paul Eberhard, who has also contributed a chapter to this book. His original emphasis was as much on what neuroscience can tell us about architecture for people who are different from us as it is about the experience of the built environment by someone like oneself[9] (which, of course, varies from self to self, and from architect to architect). How can neuroscience inform the architect designing a building for those whose experience differs greatly from her own? For example, how might one design a building to help keep the elderly alert as they move about their environment? Eberhard also considered how classroom design might affect the cognitive processes of children. In this case, he was trying to address the notion of neural plasticity, the process of ongoing change that takes place within the brain. Other related issues include work environments and their effect on productivity. We could even look at circadian rhythm and, for example, the lighting of rooms in hospitals or homes for the elderly. If the nurse has to enter the room in the middle of the night, how do we stop the light from disrupting the circadian rhythm, and thus presumably impairing the healing process or the restorative power of sleep?

One area of the brain, the hippocampus, deserves some special attention because of its role in navigation.[10] The hippocampus is also very interesting in human terms because of the case of H. M., whose identity as Henry Gustav Molaison was revealed after his death. When he was young, H. M. had uncontrollable epilepsy, and his neurosurgeon William Scoville, in Montreal, removed a huge portion of the brain centered on the hippocampus. As a result, he lost *episodic memory*, the ability to form new memories of episodes in his life, even though he could recall episodes from life before the operation. One might be talking to him, leave the room, come back a few minutes later, and apologize for the delay—but he would not remember having ever seen you. He still had *working memory*, the ability to recall items relevant to an ongoing activity, but once that activity ceased, this transient memory could no longer be activated.[11] And yet there was a surprise: if one played a new board game with him one day (remember: he could exploit working memory), the next day he would insist that he had never seen the board game. And so it would be on successive days. And yet, as he played the game each day, his skill kept improving. In other words, he retained some *procedural memory*, the ability to gain new skills, even though he had no explicit memory of the occasions on which he had built up those skills.[12] In summary, the hippocampus plays a crucial role in the creation of memory of episodes, yet the cerebral cortex is still able to hold old memories once they have been consolidated even when the hippocampus has been removed. Moreover, skill memory can be developed without a hippocampus, and involves regions like the basal ganglia and cerebellum. Consequently, the interaction between cerebral cortex, hippocampus, and these other regions has been of great interest to neuroscientists.

Studies of rats, however, emphasize the role of the hippocampus in spatial navigation. John O'Keefe and his colleagues in London[13] placed electrodes in the hippocampus of rats, and found that some cells would tend to respond not to what the rat was seeing or what the rat was doing, but to where it was in the lab. They called these "place cells." A single cell does not tell you (or the rat) all that much. However, one of the interesting properties of the brain as we get down to the cell circuit level is *population coding*. No single neuron "knows" anything with any precision. The populations of neurons encode knowledge between them. In the present example, one cell may signal "you're somewhere here" and another cell might signal "you're a little bit up" or "you're a little bit to the right"—and between them the population gives the rat very precise information about its location. Of course, one of the big mysteries for neuroscience is how we, or the rat, have the embodied experience of being in a specific place when cells are firing across a broad region of the map. A partial answer is that individual experience strengthens synapses selectively and links place neurons to neurons elsewhere in the brain that represent sensory, motor, and other experiences the rat has had when traversing the regions with which each place cell correlates.

Inspired by such studies, neuroscientists have used brain imaging to show that the hippocampus plays a role for humans in navigation as well as episodic memory. For example, in 1997 Maguire[14] showed how the right hippocampus works with other brain regions in processing spatial layouts over both long and short time courses, participating in both the encoding and the retrieval of topographical memory.

All this makes the hippocampus very interesting for the insights it can offer into both our *navigation in time*, episodic memory as studied in humans, and our *navigation in space*, as shown by the place cells which bring us down to the circuit level in the hippocampus. This is why neuroscience of the hippocampus and related brain regions becomes relevant not only to *wayfinding* as a particular component or property of a building but to the way buildings can act as repositories of human memory, and thus may be structured with special concern for populations whose memory is impaired.

NEUROMORPHIC ARCHITECTURE

Neuromorphic architecture is based on the attempt to answer the question: What if a building had a "brain" or, more accurately, a nervous system? When we study an animal we are looking at a creature whose brain and body evolved together to support its exploration of, and survival in, the surrounding environment. Complementing this is the idea that if a building had a "brain," its body would surround its world. Without discounting the importance of "green" buildings and the attendant focus on energy exchange with the environment surrounding a building, I am thinking instead of an "interactive infrastructure" of a building that, for example, might contain something cognitively equivalent to the function of the hippocampus—but whereas the hippocampus helps the animal keep track of its own navigation in the external world, the building's "hippocampus" would keep track of people within the building and perhaps communicate with them to provide a whole new, adaptive, level of human support. This may sound like science fiction, but I see it as part of the future of architecture.[15]

If the building is a cognitive entity, not just a fixed structure possibly modified by shifting the furniture around, it may become dynamically reconfigurable on the basis of its interactions with the people within it. Moreover, thinking of each piece of furniture as a form of perceptual robot rather than a static piece of equipment may help us design environments that can dynamically respond to the needs of their inhabitants. To return to the issue of meeting the needs of special populations: consider how, if an elderly infirm person is trying to get out of bed, the bed would recognize her intention and assist her. In this case the bed is in some sense a robot—despite its lack of humanoid form, it is a dedicated system with coupled sensory and motor abilities, and a degree of autonomy in its ability to respond to a person, or persons, in its environment. Some might even say that the furniture would possess a limited form of empathy in meeting the needs of its human users. We are getting very comfortable now with the idea that our smartphones have various useful apps that can do different things, but the interface is still primarily symbolic, though augmented by finger motions. What happens when the interface extends to dynamic interaction with a building and its furnishings?

THE NEUROSCIENCE OF THE DESIGN PROCESS

I do not presume to know what the neuroscience of the design process will turn out to be. Rather, I want to present ideas that may constitute part of the foundation for future work in this area, using a quote from Peter Zumthor's essay "A Way of Looking at Things" in his book *Thinking Architecture*:

When I think about architecture, *images* come into my mind.

When I design, I frequently find myself sinking into old, *half-forgotten memories*. … Yet, at the same time, I know that all is *new* and that there is no direct reference to a former work of architecture. …

Construction is the art of making a meaningful whole out of many parts. … I feel respect for the art of joining, the ability of craftsmen and engineers … the knowledge of how to make things. …

Unlike the sculptor, *I have to start with functional and technical requirements*.

The challenge [is one] of *developing a whole out of innumerable details*, out of various functions and forms, materials and dimensions. … Details … lead to an understanding of the whole of which they are an intrinsic part.[16]

4.6 Peter Zumthor, Baths at Vals, Switzerland.

One comment before we proceed: Zumthor titles his essay "A Way of Looking at Things," but these quotes involve motor imagery as much as visual imagery—looking not just at things but also at how those things could be made. This recalls the notion of "action-oriented perception" mentioned above. My son, the architect Ben Arbib, has been through Zumthor's thermal baths in Vals and has talked about the way Zumthor's design manages the visitor's movement through the space, merging from one sensory experience to another. Different sensory modalities are engaged: in one room there are fragrant flowers in the water to engage the sense of smell; elsewhere a heavy leather curtain may engage a sense of tactility and heaviness in the action of moving the curtain aside.

Reading Zumthor's experience of his design process, alongside my son's description of his personal experience in one of Zumthor's buildings, brings up an important issue for "Neuroscience for Architecture": "What is the job of neuroscience here?" Neuroscience rests on the design of repeatable experiments and the development of reasoned explanations that address a growing range of empirical data. Among its goals is to understand what different parts of the brain do during various tasks, and how they interact. What do the circuits do? How do the various factors change with development, aging, and various neurological disorders? Many neuroscientists study the underlying neurochemical details and patterns of gene expression—though I think that, for now at least, these latter details are not necessarily relevant in the conversation between architects and neuroscientists. Given all this, I do not see it as a goal for neuroscience per se to explain how Peter Zumthor designed the thermal baths at Vals. What we can do instead is to glean from biographical, even autobiographical, statements of the stages that different architects go through to ask what it would take for a human brain to mediate the set of processes involved in design. Only at the latter level can we do replicable studies of brain structure. We are not going to be able to get into the particular resonance between the biography of Zumthor and the environment of Vals and the demands of thermal baths, but we may have an understanding of what processes were involved. And this may well feed back into the educational process for architecture students. If we can say "these are the skills that are needed for exceptional design," then, even for students who are preparing for nonexceptional design, it may be useful to cultivate those skills in a way that is informed by neuroscience. So let me focus on extracts from the above quote from Zumthor, and briefly note opportunities for relevant neuroscience.

What is an image? Images can be very precise, as for example the glossy photograph of a building, or they could present some general feeling, perhaps multimodal, of certain aspects of the building. So one topic we might pursue as a neuroscience project is the variety of images. How do different parts of the brain collaborate in creating images? How do those images change? Without going into the details, we may note classics on visual perception and imagery, such as the work of Gombrich and Gregory,[17] Jeannerod's work on motor imagery,[18] and studies of imagery using human brain imaging.[19] For example, in 2013 Passingham and his colleagues were able to distinguish the brain regions involved in the retrieval of recent, contextually bound information from those involved in other processes, such as visual imagery, scene reconstruction, and self-referential processing. The intriguing challenge is to go from this region-by-region analysis to assess the intricate interactions of detailed patterns of neural activity across multiple regions, to understand how neural plasticity encodes fragments of experience, and trace how the subtly rewired brain can then respond in novel ways to various tasks and contexts—with the ability to choose stimulating contexts—which is one of the hallmarks of creativity. The hippocampus may work with the cerebral cortex to recall some episode, and other parts of the brain may work to bring some aspect of a general skill into play. "Half-forgotten memories"—some of specific moments, some of general feelings, some of procedural skills—all serve to reshape the current directions in which the imagination can lead the design process.

My earlier discussion of a hand reaching for an object provides the essential basis for the human skills of construction: the use of the hands. In particular, humans are exceptionally capable in their ability to design and use tools to extend the capacities of their hands in "making a meaningful whole out of many parts." Although tool use has been observed in other species, Krist Vaesen, in his article "The Cognitive Bases of Human Tool Use," systematically compares humans and nonhuman primates with respect to nine cognitive capacities (both social and nonsocial) considered crucial to tool use: enhanced hand-eye coordination, body schema plasticity, causal reasoning, function representation, executive control, social learning, teaching, social intelligence, and language.[20] He documents striking differences between humans and great apes in eight out of nine of these domains.[21] He also demonstrates how several of these cognitive traits help to explain our unique ability for cumulative culture, as well as the astonishing technological complexity this has produced: some traits enable high-fidelity cultural transmission, yielding preservation of traits across successive generations; while others facilitate individual learning and the introduction of new cultural variants that are necessary to incremental change.

While I do not disagree with Vaesen, I would like to shift our attention from tool to construction. Many creatures can use tools *of a specific kind* and in some cases even make them, as do New Caledonian crows. Several studies[22] conclude that the complex cognitive processes involved in tool use may have independently coevolved with large brains in several orders of corvine and passerine birds. Nonetheless, it seems to me that nest building by birds is even more impressive than their tool making. In 1995, in *Animal Architecture*,[23] Juhani Pallasmaa documented the far wider range of constructions across many species. Indeed, Hansell and Ruxton[24] urge that we should view tool behaviors as a limited subclass of construction behavior. Nest building in birds has been a key driver of habitat diversification and speciation in these groups.[25] It is thus intriguing that Stewart and his colleagues[26] show that reuse of specific nest sites by savanna chimpanzees may be a result of "niche construction"[27] through formation of good building sites within trees.

They speculate that environmental modification through construction behavior may have influenced both chimpanzee and early hominin ranging by leaving behind recognizable patterns of artifact deposition across the landscape.

However, it is important to differentiate between animal architecture as the result of genetic programs which guide the behavior of animals in constructing their habitats, and human architecture which can change across many dimensions in ways greatly removed from genetic constraints. Each of the skills that Vaesen examined develops thanks in part to the genetic changes which separate humans from apes, but are then supported by the cultural evolution of new human cultural and technological environments that allow humans to develop constructions in fresh and surprising ways.

Let us shift attention, then, from "using a tool" to the ability to deploy multiple tools to solve a problem. To join a piece of wood to the wall, I may employ a screw of sufficient length plus a screwdriver; or a nail and hammer. I may also employ a stud finder, but if I need to affix an object where there is no stud, I deploy a Rawlplug, a drill, and a hammer to prepare for the screw. For household repair, I may deploy these tools and more to solve a truly novel problem by breaking it down into subproblems for which I have routine solutions. Or I may call in a handyman, thanks to the great specialization within human society, and the social construction of monetary incentives.

Of the nine cognitive capacities Vaesen lists, only three—enhanced hand-eye coordination, body schema plasticity, and function representation—relate directly to using a tool for its intended purpose. Two—causal reasoning and executive control—relate not so much to tool use as to the more general skill of problem-solving (of which construction, with or without tool use, is a crucial subcase). The remainder—social learning, teaching, social intelligence, and language—all relate to social interaction in general or the transfer of skills in particular, whether or not they involve tool use. He argues that only one of these nine capacities, *body schema plasticity*, cannot be invoked to explain what makes human technological abilities unique, since "we share the trait with our closest relatives." However, the issue is not whether the body schema can be extended, possibly by extensive shaping as in monkeys.[28] Rather, it is (in part) the uniquely human rapidity and flexibility with which different extensions of the body schema can be deployed in some overall task, switching back and forth between using some part of the body or some part of a tool as the end-effector for the current action.[29]

In this regard, it is useful to think of language as a "meta-tool," viewing the grammar of a language not as a set of very general syntactic rules but rather as involving a large number of *constructions*, which provide tools for assembling words hierarchically to meet the communicative goals of both familiar and novel situations.[30] We express our ideas by putting together words in order to have some chance of conveying to others the ideas that we have in mind. We do not have brain-to-brain transfer, yet we can put those parts (those words) together, obeying the rules of English grammar, to make a whole. But when I utter a sentence, I may seek to express an idea that does not form easily into words, and it may only be through conversation (whether with another person or with myself; whether in words alone, or mediated by hand gestures, sketches, or appeals to the passing scene) that the original idea may become shapely and well-expressed so that I can at last feel I have truly shared my insights with others. So, too, with a building: The architect may begin to shape some overall whole without yet knowing how the details will fit together, and so much work is required to shape the design—but now in an extended version of language in which both sketches and formalized plan drawings complement the written or spoken forms. The architect's design must incorporate a deep understanding of the skills of the builder and the artisan (which, these days, may themselves be rapidly changing in tandem with new technologies)—of what different forms of construction might achieve. This, no less than ideas about how the building is to be used and what aesthetic statement it is to make, will contribute to rendering the design so explicit that the actual construction process can begin.

No matter how poetic the space in Vals may be, getting the water to the appropriate temperature, creating ways for people to be in the water, and providing changing rooms are all functional requirements that the final design must satisfy. These vary from tight to very loose constraints, and we see that Zumthor transformed them in a way that no one else would have done. The conversation between "Here is what I want to realize with the building" and the final result that calls upon the whole of the architect's experience is accomplished by satisfying the functions in a way, as Zumthor says, "that is not defined by those functional requirements alone."

Note that it is not that one fixes on a detail, adds another detail, adds yet another detail, and finally the whole emerges. Rather, one has this general idea of the whole, and details begin to come into place, and to replace that vague understanding of the whole with a more precise understanding of the parts of the whole, which now constrain how you will fill in other parts of that whole.

HANDS, MIRROR NEURONS, AND THE BEAUTY OF ABSTRACTION

In my earlier discussion of control of reach-to-grasp actions, I mentioned that Rizzolatti and his colleagues in Parma recorded the activity of single neurons in area F5 of the brain of macaque monkeys and found cells that fired preferentially as the monkey executed some types of action, but not others. Years later, in a *subarea* of F5, they found cells that had a surprising extra property—such a cell fired not only when the monkey performed its preferred action, but also when the monkey observed the human experimenter perform a similar action. They called such cells *mirror* neurons[31]—neurons for which the effective observed movement is similar to the effective executed movement.

It is important to understand, as I pointed out above, that a population of cells codes a range of possibilities. It is not that one cell says "I am specific to a precision pinch," and another cell says "If I fire it means there is a power grasp." Rather, they belong to a *whole* group of related neurons. Each neuron is active for a feature of the executed or observed action. Many neurons may be active for a range of features related to the given action, but in different ways. Thus the activated population can give a nuanced representation of the current grasp, whether executed or observed—not a simple yes-no judgment. Harry Mallgrave suggests that mirror neurons and embodiment relate empathy for a person to empathy for a building. The implication, then, is not to think of empathy in an all-or-none way but, rather, to think of how a diverse population of neurons might fire in different circumstances. If enough of the population fires in relation to the building, it might trigger a related population that corresponds to your interaction with another person. In other cases, a different subpopulation could be active, and one's reaction to the building might then be more abstract rather than empathic.

I want to emphasize that mirror neurons are just a small part of the brain. They do not perform action execution and recognition (or empathy) all by themselves. Consider an fMRI (brain imaging) study of humans recognizing actions performed by humans as well as other species.[32] When the subject being scanned observed, at separate times, videos of a human biting, a monkey biting, and a dog biting, an area of the subject's brain thought to contain mirror neurons for orofacial movements was active in all three cases. In this case, the mirror neurons do seem to be active in mediating the subject's understanding of *biting*. However, the story becomes more complicated when the general concept is *oral communication*. Now the subject observed separate videos of a human talking (no soundtrack, but with lip movements), a monkey chattering its teeth or smacking its lips, and a dog barking. There was a lot of "mirror system" activation while observing the human, a small amount of activation while observing the monkey, and none while observing the dog. Of course the subject recognizes that the dog is barking, but mapping onto mirror neurons is not part of that particular understanding. Thus, where Buccino and his team say: "Actions belonging to the motor repertoire of the observer are mapped onto the motor system, in particular the mirror neurons. Actions that do not belong to this repertoire are recognized without such mappings making a strict dichotomy between what you can do and what you can't do,"[33] I would say that all these actions can be recognized without the aid of mirror neurons, but if an action is in our own repertoire, the mirror neuron activity enriches it by tying it in to our own motor experience.

Where Juhani Pallasmaa's book *The Thinking Hand* emphasized the role of the hand in, for example, invoking our embodied experience in drawing sketches for further design, I have used my understanding of mirror neurons to develop a theory of the evolution of the human language-ready brain. My book *How the Brain Got Language* shows how the brain mediates a conversation between vision and the hand not only in praxis but also in communication, including language.[34] In particular, where both Pallasmaa and Mallgrave and others in this volume emphasize embodiment, I want to say that rationality may to a greater or lesser extent be disembodied, in the sense that language and culture support abstract ways of thinking, so that (some of us) can come to understand ideas such as debt, justice, the Higgs boson or infinity, whose links to embodied experience may be tenuous.[35] This power for abstract thinking is immensely important to us as humans, so long as it does not make us neglect our embodied interaction with the social and physical (including built) environment. So, returning to the idea of wayfinding, as architects one might assess, on the one hand, how people will experience their passage through the spaces of a building based on cues afforded by the spatial structure of that building and, on the other hand, how to use signage, appealing to the symbolic aspect of their cognition to help assist their navigation when other cues fail.

A CASE STUDY IN GROUP CREATIVITY: CHOREOGRAPHY

If we are to continue the dialogue between architecture and neuroscience, it will be important to extract case studies of architects engaged in design. (I am sure there are many case studies in existence, but I have not looked at them yet.) Meanwhile, it is instructive to consider case studies of group creativity in other disciplines. As I have said, I do not think we can expect neuroscience to shed much insight into the autobiographical particularities in a specific case study, but we may begin to see that there are certain skills that are crucial to design, and then discern different styles. Certain people might excel at visualization, others may rely on tactile images, while yet other people may skillfully imagine moving through a building and creating three-dimensional patterns of interaction. However, in lieu of such studies, I am going to examine the design of a dance, in particular a case that involved group creativity: although the choreographer controlled the overall development, interactions with and between the dancers played a crucial role. The dance was created by Anna Smith as choreographer working with eight female dancers in Melbourne, Australia. In 2005, Kate Stevens gave a chronology of the creation of the dance in the rather provocatively titled book *Thinking in Four Dimensions: Creativity and Cognition in Contemporary Dance.*[36] I offer a brief encapsulation of what it took to create that dance, in order to suggest how a careful case study of the creative process may offer clues for scholars who want to know what a human brain has to be in order to make such processes possible. Then you can ask the neuroscientists to tell you more about how those processes work in the brain, and you can ask more of the case study to determine what happened in this particular instance.

When I am writing, I keep multiple drafts of the document, which serve as my record of this particular creative process. And many architects, when they are creating, make sketches, either by hand or on the computer—and these provide the cumulative record of the various ins and outs of the design process. But a dance is evanescent. In the present case, a journal was kept with a record of what ideas were being explored in each particular week of the dance. This included photos of, say, a particular pose, but even more important were the videos of particular patterns of movement, creating a sort of external memory of the development. I am reminded of the work of Merlin Donald,[37] who has a theory of human evolution based on mimesis. For him, the third and perhaps most crucial stage of development is the transition from reliance on memory structures to external devices, whether through sketching, drawing, building sculptures, or what have you. For this dance, a new dimension of technology (the video) helped produce a record in which patterns of movement were essential.

With much less constraint than that of Peter Zumthor at Vals, the initial idea for the dance was to think about the color red and its relation to blood while having the dancers come up with various "through-lines," ways of moving out through the body. And through the video record as well as through observations of each other, the group built up a motion vocabulary from which one could move out of the body toward this general idea of *redness* and blood. In the second week, they began to systematize those movements out from the body that they thought were important. In that same week, a turning point came when the choreographer asked each of the dancers to bring something red to rehearsal. One of them brought a bag of red kidney beans. Somehow the mixture of the feel, the sound, the texture and the pouring of the kidney beans became a central metaphor for the idea of blood, and this metaphor began to take control of the group's imagination.

As the dance developed, the idea of pooling blood around a dancer provided a powerful image (figures 4.7 and 4.8). Moreover, pouring the kidney beans out on the floor had an auditory component that was reminiscent of rain on a tin roof, inspiring the name *Red Rain* for the dance. Outlining the shape of the body with the beans, and then having the dancer move, formed various interesting outlines that added to the emerging motions of the dance.

4.7 AND 4.8 Red kidney beans were used to outline the shape of a prone body. The line of the body left behind suggested traces written in blood.

In Weeks 1 and 2, the focus was on arterial blood, red blood. What about blue, venous blood? This question led, in Week 3, to the creation of large blue and red objects, symbolizing the two forms of blood, as props. And then the interaction of one dancer with those props yielded another beautiful, powerful image, the "nesting child," that became another building block for the dance. We see again this open-ended variation on what has been done before, but we also see this "ratcheting"[38] where every now and again something will crystallize as a new component for the emerging design. Creativity exploits the prior schemas, places them in new contexts, and assembles them in new ways to transform them into novel schemas which—together with external props and "narrative momentum"—provide aliment for further variation and development, building on what you know in order to vary it and discover something new. What one thing means at the beginning of a performance may change completely as the performance unfolds.

By Week 5, a lot of the through-lines had been created, but there was concern about stasis, so the choreographer forced new variations on the existing themes by getting the dancers, much to their discomfort, to truncate their through-lines and learn new ones. In this way, a new set of bodily movements emerged which were perhaps less natural but more evocative. And so it went on. The new through-lines, which at first had been hard to master because it was difficult to move the body in these unnatural ways, became automatized, allowing the dance to move up a level to examine how to sequence and interrelate the design. And so, over the remaining weeks, the dance became complete. What may resonate with many architects is the frustration that the choreographer, Anna Smith, felt at times, not knowing what the shape of the final work would be because it was an emergent process rather than a predefined product. Creativity in composing this dance involved sequencing, melding, and linking the parts of the work as well as the creation of the parts themselves.

This takes us back to a final visit with Peter Zumthor. His essay "A Way of Looking at Things" led us to ponder a variety of words and phrases relevant to the design process, and thus helped us begin to delineate some of the challenges for a neuroscience of the design process.

Images
Half-forgotten memories
Yet all is new
Construction is the art of making a meaningful whole out of many parts
I have to start with functional and technical requirements
Developing a whole out of innumerable details.

We now have some sense of how images function and how memories are built up and recorded (both internally and externally), as well as the continual search for something new. In terms of construction, what is not emphasized in the quote from Zumthor—but what I have come back to several times—is that the whole is creating the parts at the same time that it is being created from the parts. Part and whole do not form an either/or. In general, we neither start with a whole and fill in the details, nor do we start with a known set of parts and merely assemble them. Instead there is a continual dialogue between the emerging selection and innovation of parts and the emerging concept of the whole. Indeed, progress in cognitive neuroscience emerges as we not only work down from psychological terminology in search of brain mechanisms but also work up from an understanding of neurons to enrich our understanding of the functional possibilities of neural circuitry. I have, among many other topics, touched briefly on the schema theory of both action and visual perception to sketch the essence of cooperative computation in which diverse schemas must compete and cooperate before a coherent understanding may emerge. I hold this to be the essential style of the brain. And the notion of dialogue is crucial—not only between schemas or parts of the brain, or between architect and sketchpad, but also between the many collaborators who may help shape the design, whether through a specific contribution to the emerging whole, or through informed critique. We begin to see, dimly, what enables the human brain to design new constructions, whether in our everyday life, or the choreography of a dance, or the development of a building.

Bringing the knowledge base of neuroscience to the theory and practice of architecture will not be "plug and play." The Society for Neuroscience has annual meetings that attract around 30,000 neuroscientists, and there are some 15,000 presentations, both talks and posters. However, virtually none of these presentations reports studies designed explicitly to meet the challenges of architecture. Progress will come only if *some* architects and *some* neuroscientists learn to work together to match new goals to new methods of research—using issues in architecture to define the new experiments in neuroscience. This is a call for collaboration, not a case of neuroscientists being already able to answer all the architects' questions—though I believe neuroscience already has a range of insights that are new to architecture and can stimulate its future development.

NOTES

This chapter is based primarily on a presentation given on June 3, 2013, in Helsinki as part of the symposium "Architecture and Neuroscience," the first in a series celebrating Tapio Wirkkala's 100th birthday in 2015, organized by Aalto University, Alvar Aalto Academy, Finnish Center for Architecture, University of Helsinki, and Tapio Wirkkala–Rut Bryk Foundation. My thanks to Juhani Pallasmaa and Esa Laaksonen for conceiving this event and bringing it to fruition. The chapter also incorporates material from my presentation in November 2012 in Taliesin West at the symposium "Minding Design: Neuroscience, Design Education and the Imagination," organized by Sarah Robinson for the Frank Lloyd Wright Foundation, and with the support of the Academy of Neuroscience for Architecture.

1. In this chapter, the word *architecture* will almost always be used in the sense of "the architecture of the built environment." However, as we shall see briefly, neuroscientists use expressions like *neural architecture* or the *architecture of the brain* to describe the patterns whereby neurons and their connections are arranged in layers, columns, and regions in the three-dimensional structure of a brain. A computer scientist will then use the term *neuromorphic architecture* for a computer which, unlike a

conventional serial computer, has its components laid out in a fashion inspired in part (often a very small part) by the architecture of the brain.

2. Michael Arbib, *How the Brain Got Language: The Mirror Systems Hypothesis* (New York: Oxford University Press, 2012), chapters 4 and 5.

3. Michael Arbib, *The Metaphorical Brain: An Introduction to Cybernetics as Artificial Intelligence and Brain Theory* (New York: Wiley-Interscience, 1972).

4. Michael Arbib, *The Metaphorical Brain 2: Neural Networks and Beyond* (New York: Wiley-Interscience, 1989).

5. Jean Piaget, *The Construction of Reality in the Child* (New York: Norton, 1954); Jean Piaget, *Biology and Knowledge: An Essay on the Relations between Organic Regulations and Cognitive Processes* (Edinburgh: Edinburgh University Press, 1971).

6. Michael Arbib, "Rana Computatrix to Human Language: Towards a Computational Neuroethology of Language Evolution," *Philosophical Transactions of the Royal Society of London, Series A Math, Physics, Engineering Science* 361 (2003): 2345–2379.

7. Marc Jeannerod, Michael Arbib, Giacomo Rizzolatti, and H. Sakata, "Grasping Objects: The Cortical Mechanisms of Visuomotor Transformation," *Trends in Neuroscience* 18 (1995): 314–320.

8. Sarah Robinson, *Nesting: Body, Dwelling, Mind* (Richmond, CA: William Stout, 2011).

9. John P. Eberhard, *Brain Landscape: The Coexistence of Neuroscience and Architecture* (Oxford: Oxford University Press, 2008).

10. Esther Sternberg and M. A. Wilson, "Neuroscience and Architecture: Seeking Common Ground," *Cell* 127 (2006): 239–242.

11. William Scoville and B. Milner, "Loss of Recent Memory after Bilateral Hippocampal Lesions," *Journal of Neurology, Neurosurgery and Psychiatry* 20 (1957): 11–21; rpt. *Journal of Neuropsychiatry and Clinical Neurosciences* 12 (2000): 103–113.

12. B. Milner, S. Corkin, and H. L. Teuber, "Further Analysis of the Hippocampal Amnesic Syndrome: 14-Year Follow-up Study of H. M.," *Neuropsychologia* 6 (1957): 215–234.

13. John O'Keefe and J. O. Dostrovsky, "The Hippocampus as a Spatial Map: Preliminary Evidence from Unit Activity in the Freely Moving Rat," *Brain Research* 34 (1971): 171–175.

14. E. A. Maguire, "Hippocampal Involvement in Human Topographical Memory: Evidence from Functional Imaging," *Philosophical Transactions of the Royal Society of London. Series B: Biological Sciences* 352 (1997): 1475–1480.

15. Michael Arbib, "Brains, Machines and Buildings: Towards a Neuromorphic Architecture," *Intelligent Buildings International* 4 (2012): 147–168.

16. Peter Zumthor, *Thinking Architecture*, 3rd edn. (Basel: Birkhäuser, 2012); emphasis added.

17. E. H. Gombrich and R. L. Gregory, eds., *Illusion in Nature and Art* (London: Duckworth, 1980).

18. Marc Jeannerod, "The Representing Brain: Neural Correlates of Motor Intention and Imagery," *Behavioral and Brain Sciences* 17 (1994): 187–245.

19. G. Ganis, W. L. Thompson, and S. M. Kosslyn, "Brain Areas Underlying Visual Mental Imagery and Visual Perception: An fMRI Study," *Cognitive Brain Research* 20 (2004): 226–241; A. Ishai, L. G. Ungerleider, and J. V. Haxby, "Distributed Neural Systems for the Generation of Visual Images," *Neuron* 28 (2000): 979–990; R. E. Passingham, J. B. Rowe, and K. Sakai, "Has Brain Imaging Discovered Anything New about How the Brain Works?," *Neuroimage* 66 (2013): 142–150.

20. Krist Vaesen, "The Cognitive Bases of Human Tool Use," *Behavioral and Brain Sciences* 35 (2012): 203–218.

21. The present section is based, in great part, on my commentary "Tool Use and Constructions," *Behavioral and Brain Sciences* 35 (2012): 218–219.

22. G. R. Hunt, "Manufacture and Use of Hook-Tools by New Caledonian Crows," *Nature* 379 (1996): 249–251; A. S. Weir, J. Chappell, and A. Kacelnik, "Shaping of Hooks in New Caledonian Crows," *Science* 297 (2002): 981; L. Lefebvre, N. Nicolakakis, and D. Boire, "Tools and Brains in Birds," *Behavior* 139 (2002): 939–973.

23. Juhani Pallasmaa, ed., *Eläinten arkkitehtuuri/Animal Architecture* (Helsinki: Museum of Finnish Architecture, 1995).

24. Mike Hansell and Graeme Ruxton, "Setting Tool Use within the Context of Animal Construction Behaviour," *Trends in Ecology and Evolution* 23 (2008): 73–78.

25. N. E. Collias, "On the Origin and Evolution of Nest Building by Passerine Birds," *Condor* 99 (1997): 253–269; Mike Hansell, *Bird Nests and Construction Behavior* (Cambridge: Cambridge University Press, 2000).

26. F. A. Stewart, A. K. Piel, and W. C. McGrew, "Living Archaeology: Artefacts of Specific Nest Site Fidelity in Wild Chimpanzees," *Journal of Human Evolution* 61 (2011): 388–395.

27. A. Iriki and M. Taoka, "Triadic Niche Construction: A Scenario of Human Brain Evolution Extrapolating Tool-Use and Language from Control of the Reaching Actions," *Philosophical Transactions of the Royal Society B: Biological Sciences* 366 (2011); K. N. Laland, F. J. Odling-Smee, and M. W. Feldman, "Niche Construction, Biological Evolution, and Cultural Change," *Behavioral and Brain Sciences* 23 (2000): 131–146.

28. Ishai, Ungerleider, and Haxby, "Distributed Neural Systems for the Generation of Visual Images"; M. A. Umiltà, L. Escola, I. Intskirveli, F. Grammont, M. Rochat, et al., "When Pliers Become Fingers in the Monkey Motor System," *Proceedings of the National Academy of Sciences* 105 (2008): 2209–2213.

29. Michael Arbib, J. J. Bonaiuto, S. Jacobs, and S. H. Frey, "Tool Use and the Distalization of the End-Effector," *Psychological Research* 73 (2008): 441–462.

30. A warning about terminology: in *construction grammar* (M. A. Arbib and J. Y. Lee, "Describing Visual Scenes: Towards a Neurolinguistics Based on Construction Grammar," *Brain Research* 1225 [2008]: 146–162; W. Croft, *Radical Construction Grammar: Syntactic Theory in Typological Perspective* [Oxford: Oxford University Press, 2001]; A. E. Goldberg, "Constructions: A New Theoretical Approach to Language," *Trends in Cognitive Science* 7 [2003]: 219–224; D. Kemmerer, "Action Verbs, Argument Structure Constructions, and the Mirror Neuron System," in M. A. Arbib, ed., *Action to Language via the Mirror Neuron System* [Cambridge: Cambridge University Press, 2006], 347–373; A. Verhagen, *Constructions of Intersubjectivity: Discourse, Syntax, and Cognition* [Oxford: Oxford University Press, 2005]), the term *construction* is used for a basic "assembly rule" for combining words or phrases into a larger entity—for example, adding an adjective to a noun phrase to make a larger noun phrase, as in "wilted + red rose → wilted red rose"—while keeping track of how both the form and the meaning of the components are combined. Thus a sentence like this one is the result of applying a whole hierarchy of constructions, whereas our everyday usage might suggest that it is the sentence that is the construction, and that what linguists call constructions are "assembly rules."

31. G. di Pellegrino, L. Fadiga, L. Fogassi, V. Gallese, and G. Rizzolatti, "Understanding Motor Events: A Neurophysiological Study," *Experimental Brain Research* 91 (1992): 176–180; V. Gallese, L. Fadiga, L. Fogassi, and G. Rizzolatti, "Action Recognition in the Premotor Cortex," *Brain* 119 (1996):

593–609; G. Rizzolatti and L. Craighero, "The Mirror-Neuron System," *Annual Review of Neuroscience* 27 (2004): 169–192; G. Rizzolatti and C. Sinigaglia, *Mirrors in the Brain: How Our Minds Share Actions, Emotions, and Experience*, trans. Frances Anderson (Oxford: Oxford University Press, 2008).

32. G. Buccino, F. Lui, N. Canessa, I. Patteri, G. Lagravinese, et al., "Neural Circuits Involved in the Recognition of Actions Performed by Nonconspecifics: An fMRI Study," *Journal of Cognitive Neuroscience* 16 (2004): 114–126.

33. Ibid.

34. Inspired by Pallasmaa's *The Thinking Hand*, but eager to show how our ability for abstract thought complements our ability to draw on our embodied interactions with the world around us, I gave the title "From Hand to Symbol and Back Again" to my talk at Taliesin West (Juhani Pallasmaa, *The Thinking Hand* (London: Wiley, 2009).

35. Michael Arbib, Brad Gasser, and Victor Barrès, "Language Is Handy but Is It Embodied?," *Neuropsychologia*, vol. 55 (March 2014): 57–70.

36. Kate Stevens, "Chronology of Creating a Dance: Anna Smith's *Red Rain*," in R. Grove, Kate Stevens, and S. McKechnie, eds., *Thinking in Four Dimensions: Creativity and Cognition in Contemporary Dance* (Melbourne: Melbourne University Press, 2005), 169–187.

37. Merlin Donald, *Origins of the Modern Mind: Three Stages in the Evolution of Culture and Cognition* (Cambridge, MA: Harvard University Press, 1991).

38. C. Tennie, J. Call, and M. Tomasello, "Ratcheting Up the Ratchet: On the Evolution of Cumulative Culture," *Philosophical Transactions of the Royal Society of London, Section B, Biological Sciences* 364 (2009): 2405–2415.

5

TENDING TO THE WORLD

Iain McGilchrist

Many scientists assume that describing something at the brain level reveals the ultimate truth about its nature. However, as Wittgenstein, among others, observed, nothing can ever be *reduced* to anything: it is what it is. People got terribly excited when neuroscientists found a brain circuit that "lit up" when you fell in love; perhaps you remember the splash it made in the papers. We were invited to think that this brain activity told us something about the business of falling in love, just as describing spiritual experience at the brain level seems to have explained that experience away. Of course it has done nothing of the kind; something is bound to be going on in the brain when we do anything—or nothing—at all. Rather than probing the brain for some ultimate truth, I want instead to discuss the central role that attention plays in constituting the world, and to explore how the brain helps to shape and limit the different kinds of attention we are able to pay.

PAYING ATTENTION

The way in which we attend to the world governs what we find in that world. Behind my house in Scotland there is a large mountain which is visible from the sea: the name of the place, Talisker, comes from a Norse word meaning "sloping rock." Thousands of years ago, when the Norsemen came down to explore those parts, they saw it as a landmark to steer by. To the indigenous people it also had a meaning: for them it was the home of the gods. Since the nineteenth century, people have come to paint and photograph it: for them it is a many-textured form of beauty. Others want to mine it for the stone that it yields, and to them it is a prospect for wealth. The geologist's description of its basaltic columns is no more "real" than the others. And to the physicist, the mountain is nothing material at all—just an aggregate of probabilities of particles so small that we do not know what they are. What I want to suggest is that there is no "real" mountain apart from our many conceptions of it. The world that exists is the world we know, and our knowing comes into the experience and the creation of that world. Attention is a deeply creative act.

Whatever it is that your attention first yields governs the kind of further attention you pay, which in turn ensures you will see more of the same; and so things tend to firm up on the basis of where you started—which was just one of many possible choices. This, then, becomes your take on the world. We all have a take. It is impossible not to have a particular, partial take. When you try to cut the particular out, you still have a take, and a partial one—it just turns out to be the least human one. The models we use to understand the world influence what we find in that world. When we say we understand something, what we mean is that it resembles something else in our particular model of the world, which we think we already understand. Our knowledge is the result of this recursive process that confirms and reconfirms itself; this makes the model we choose extremely important. Too often, we take for granted the model of the machine, whereas until recently we modeled phenomena according to living, changing, organic images: the family, the tree, the river, and so on.

THE DIVIDED HEMISPHERES

The argument of my book *The Master and His Emissary* is complex and its subject matter wide-ranging, and it is supported by extensive references to scientific literature that I cannot offer here.[1] All I can do is to provide a few headlines, and hope that if you want evidence you will go to the book itself. So with that caveat, let me attempt to explain my central argument, which concerns this all-important matter of attention.

My interest began with a question so obvious I had never heard it asked at all: if the whole purpose of the brain is to make connections, why does it have a whopping divide down its middle? On the face of it, this is a waste of "computing" power: evolution could have converted it into one single mass, but it has not. The plot thickens when you realize that the corpus callosum, the body of tissue that connects the hemispheres, far from enlarging over the course of evolution, has actually reduced in relative size.[2] And, what is more, much of its function is inhibitory in nature, effectively telling the other hemisphere to "keep out."[3] Why should that be?

There are also objective, measurable differences between the two hemispheres. They have different sizes and shapes,[4] different gyral conformations on the surface,[5] different gray to white matter ratios,[6] different cytoarchitecture in places;[7] they exhibit different responses to neuroendocrine hormones,[8] and rely on different proportions of neurotransmitters.[9] Added to which, any neurologist could tell you that there are reliable differences in what happens to the subject's world depending on the side of a cerebral lesion. What is this careful separation of the hemispheres for, and why all of these differences?

In the 1960s and 1970s, an idea got about that language and reason were located in the left hemisphere and emotion and creativity in the right. Unsurprisingly, this crude notion turned out to be false. The discovery that both hemispheres are involved in every single human activity has even led some neuroscientists to deny any difference between them. They were compelled to this illogical position by thinking of the brain as a machine, and asking, as one would ask of a machine, *what it does*. But if instead they had thought of the brain hemispheres as part of a person, they might have asked the sort of question that we might ask of humans—how, *in what way*, in what manner, to what end and with what values, did they do what they did? If you ask this question, you find some fascinating and quite consistent differences.

Other animals, including birds and fish, have divided hemispheres. While human neuroscientists were blinded by the dogma of no difference, students of animal behavior were quietly indulging in what scientists are supposed to do: patiently observing what was actually happening. They found that birds and other animals use the two halves of their brain reliably quite differently.[10] They found that lateralization, and above all hemispheric differentiation, is an aid to survival, and that animals not properly lateralized tended to be at a disadvantage.[11] Why would that be?

In order for a bird to survive, it has to be able to manipulate objects—to pick up a twig to build a nest, to pick out that small seed on the background of pebbles or gravel on which it lies. But if the bird focused only narrowly on what it was already interested in, it would not last long. While it was getting its lunch, it would become somebody else's. Survival depends on paying two kinds of attention at once: one kind is narrowly focused and committed to a single end, while the other is broad, open, sustained and vigilant for whatever might be, without preconception.[12] Simultaneously paying two opposed kinds of attention is very hard to do with one brain. Its bihemispheric structure appears to have been the solution to this problem.

This adaptation naturally persisted in human beings. Neurologists conventionally distinguish five different types of attention; three and a half of these are served by the right hemisphere and one and a half by the left. The essential difference is that focused, narrow attention is the prerogative of the left hemisphere and broad, sustained attention is that of the right.[13] People who suffer damage to the right hemisphere develop a pathological narrowing of the window of attention.[14] Since attention changes the world, this means that the two hemispheres underwrite two kinds of being in the world. In ordinary daily living we alternate between or merge these different kinds of attention at a level below consciousness. Yet the implications are significant beyond measure.

TWO MINDS—TWO WORLDS

The right hemisphere understands the whole not simply as the result of assembling a bunch of fragments, but rather as an entity prior to the existence of the fragments.[15] There is a natural hierarchy of attention, global attention coming first.[16] Normally when you see, you first see an "H" and a "4" and only later see the "E's" and "8's." If you see a Dalmatian dog, you can't deduce this from the parts that compose it, which appear to be just a random assortment of black and white splashes and splodges. You have to see it as a whole first.

The left hemisphere, because it isolates things in a very narrow attentional field, tends to see them out of context. The right hemisphere sees a broader field in which things are connected, so it sees things in context, the way they are actually situated in the world. As John Dewey observed, the gravest mistake of philosophy is to remove things from their context in order to understand them.[17]

To generalize and extract where we need to particularize and see the whole in context changes the experience of space and time. So it is that the right hemisphere does not simply see a succession of points in space or time, but rather a continuum.[18] The right hemisphere is better at understanding depth and extension of time and space, and for

```
EEEE          EEEE                 88888888
EEEE          EEEE              88888 88888
EEEE          EEEE             88888  88888
EEEE          EEEE            888888  88888
EEEE          EEEE           888888   88888
EEEEEEEEEEEEEEEEEE          888888    88888
EEEEEEEEEEEEEEEEEE      8888888888888888888888
EEEE          EEEE      8888888888888888888888
EEEE          EEEE                    88888
EEEE          EEEE                    88888
EEEE          EEEE                    88888
EEEE          EEEE                    88888
EEEE          EEEE                    88888
```

5.1 Normally you first see an "H" and a "4," and only later the "E's" and "8's."

5.2 You see a Dalmatian dog, but you cannot deduce this from the parts that compose it, which appear to be just a random assortment of black and white splashes and splodges. You see it as a whole first.

that reason is responsible for the creation or interpretation of perspective.[19] Instead of observing the world as if it were on a flat screen, the right hemisphere connects with the world and understands depth in space, and harmony (its equivalent) in music.[20]

Partly because most things come into awareness from the periphery of the attentional field, the right hemisphere is better adapted to deal with new information.[21] The famous neuroscientist Elkhonon Goldberg spent much of his career demonstrating that, in fact, the right hemisphere processes things when they are fresh and new, regardless of what they are.[22] An idea, an image, a person or a sound, if new, tends to be better apprehended in the right hemisphere. When the information becomes familiar it shifts over to be categorized by the left hemisphere, which is concerned to narrow things down to the familiar and certain. And certainty—as you know from modern physics, and no doubt just from living a life—is a rare commodity in this world. Yet to interact in the world, we need the illusion of certainty, and that is the concern of the left hemisphere.[23] To achieve this sense of certainty, the left tends to narrow things down to "either/or" and "black or white."

Certainty requires clarity, and clarity, too, is an illusion. Ruskin pointed out that, from a quarter of a mile away, a white square on a lawn could be a white handkerchief or an open book. As you get closer, you see it is a book, not a handkerchief, but you cannot read the words. As you get closer still, you read the text, but then you look more closely at the page, and notice that it has ridges. Put the ridges under a microscope and you find tiny fibers. Now you could carry on this process with an electron microscope until you found not much of anything at all. At which point, do you see it clearly? Clarity, it seems, is not a characteristic of a type of perception. Clarity is a characteristic of a type of knowledge, the type offered by the categorizing left hemisphere when it is able to say: "It is one of those."[24]

The left hemisphere renders things explicit.[25] Explicitness is not always helpful: in fact it can destroy meaning. From my experience of trying to teach literature in my youth came a book called *Against Criticism.*[26] My thesis was that the whole process of criticism works against that of art. The artist took trouble to make, from the stuff of life, something absolutely unique, instantiated just in the form that it was. In this sense the work of art is embodied, incarnate—it is as unique as any of your friends. You would not make abstract generalizations about a person if you wanted to convey an idea about that person to someone else; you would instead have to introduce the person. Works of art have to be experienced, and if the whole process of criticism is to render the specific general, the concrete abstract, and the incarnate disembodied, then the work crumbles, leaving you with nothing but a handful of dust. Making poems explicit kills them, as it kills a joke.

The right hemisphere deals with the metaphoric aspects of language.[27] All that is implicit comes to us despite, not because of, the direct meaning of the words as they might be put together by a computer wielding a dictionary. Meaning comes through all that one learns from living: tone of voice, humor, irony, facial expression, the language of the body. The right hemisphere takes a broad view so that it can constitute the whole world to which we belong, whereas the left hemisphere, with its interest in manipulating the world and grasping it, constitutes only part of a world, that part of space to the right of us that it can control, the part where you literally grasp things with your hand. The left hemisphere's knowledge is the aspect of knowing described by the phrase "I've grasped it." Which really means that you have put it into the box of familiar things you understand.

In keeping with its particulate way of looking at things, the left hemisphere prefers fixity to flow. It sees a sequence as made up of static points, like a cine film made of successive frames. In the condition known as *palinopsia*, which typically results from right posterior cortical damage, the person sees the world as a sequence of stills rather than a visual flow.[28] Of course, there is a profound difference, philosophically speaking, between stasis and flow. When you start breaking down the flow into static moments, you run into all sorts of impossibilities—the stuff of Zeno's paradoxes.

The left hemisphere operates according to a mechanistic model—its world is an assemblage of bits and pieces. Le Corbusier conceptualized the house as a "machine for living" using the same model. But ultimately, as Sarah Robinson points out, Corbusier abandoned the machine for an organic model of the world.[29]

Descartes, too, moved from the dogmatic and classically Cartesian view of the *Discourse* to a more human position in his last writings, *Les Passions de l'Âme*. The later Wittgenstein of the *Investigations* is different from the young Wittgenstein of the *Tractatus*. One can see all these progressions as progressions from the left hemisphere's mechanical, rationalistic model of the world to a more complex, nuanced, subtle, interactive model of the world. The neuroscientist V. S. Ramachandran calls the right hemisphere the devil's advocate,[30] because it is always saying: "Look, before you categorize this, it might actually be something quite new, original, special, different, unique." As Heidegger would have said, the right hemisphere sees the world as it "presences," or presents itself to us, before we have processed it, before we have represented (literally re-"presented") it in our minds.

So these two minds, these two worlds, that the hemispheres underwrite, have quite different qualities. One, that of the left hemisphere, is made up of disembodied, abstract, fixed, static, discrete entities that are familiar, and general in nature. They can be put together to form a world we think we understand and can control. But this world is self-enclosed, and lifeless, compared with the world of the right hemisphere, where everything is new, interconnected, incarnate, flowing, evolving, and unique.

We differ from other animals by having highly developed frontal lobes—the most recent part of the brain to evolve. Even in an intelligent animal such as the dog, the frontal lobes constitute no more than 7 percent of the brain, and in lesser apes they constitute about 17 percent, but in the great apes and ourselves it is about 35 percent. This difference in proportion is very important for the sort of world in which we live, because the frontal lobes are dedicated to stopping us from reacting too immediately to things. They put distance between us and the world, in space and time, and endow us with the ability to stand back and think. This distance gives us the chance to question whether others might have a shared interest with ourselves, whether we might form an alliance and cooperate with them, rather than immediately bite back. They also enable us to consider the future consequences of actions: they are therefore essential to decision-making, and planning, as well as understanding other minds.

Each hemisphere has its frontal lobe: on the left it enables us to outmanipulate, which is why neuroscientists call the human mind Machiavellian. But there is another kind of intelligence, which I call the Erasmian intelligence. Around the same time that Machiavelli was writing *The Prince*, another greater thinker, Erasmus, was writing a book called *The Education of a Christian Prince*. Where Machiavelli provided the short-term advice—to be feared rather than loved—Erasmus suggested it was better for a prince to be loved rather than feared. The right frontal cortex of the human brain is larger than the left frontal cortex, because its expansion has enabled the development of social intelligence—allowing us to cooperate, to empathize, and to read the complex and subtle messages that comprise most of our communication.

The frontal cortex inhibits the posterior cortex; the more recently evolved cortex inhibits the lower, more primitive subcortical centers; and the hemispheres engage in a mutually inhibitory relationship, allowing each to better perform its task. Negation can be creative; denying one thing allows something else to stand forth. We do not assemble a sculpture from bits and pieces; we clear away the rock around it, allowing the sculpture to stand forth. This is how we discover truth, *aletheia* (literally, unconcealing) as Heidegger called it—truth is disclosed in the act of uncovering something. This clearing away to reveal something better is central to all true acts of creation. It is in acts of criticism, too. Unfortunately, some critics tend to get between you and their subject, but their real job is to clear the view so that the artist speaks for himself or herself.

So on the one hand, you have a hemisphere whose purpose, if you like, is to grasp, to manipulate the world. And on the other, you have a hemisphere whose ultimate goal is to relate to the world and understand it. The provision of what I call "necessary distance," of standing back, provided by the frontal lobes, enables us to relate better. When you're close up to something, you can't read it. When it's too far away, you can't read it either. There is a proper distance for interaction with the world, just as there is a proper distance in relationships (couples can be too "fused" or too distant). The Latin word *tendere*, the root of the word "attention," means to reach out a hand. This can be for grasping, the aim of the left hemisphere, or for exploring the world, to understand and connect—the aim of the right.

A bias toward one kind of attention over another has cultural consequences. Although all of our activities involve both hemispheres, in practice we tend to favor one take on the world. And these takes are not strictly compatible, which is why they have to be separated. They are both necessary: a balanced person takes them both, alternately or together in some way, "both/and," something which the right hemisphere would appear to understand. But the left hemisphere is "either/or" in its approach, and this exclusivity underlies the idea of the Master and his emissary, which I borrowed loosely from Nietzsche. According to this fable, a spiritual Master sent off his brightest emissary to do certain kinds of work on his behalf, because he, the Master, knew he could not get involved without losing his overall vision. The emissary, not knowing what it was he didn't know, thought that he was the doing all the important work, and failed to report back to the Master. Indeed he put himself forward *as* the Master, but once he adopted the Master's cloak, things fell apart.

We need both kinds of knowing, working together. And just as a person may tend toward one view or the other, so may a culture. A culture is, after all, an aggregate of the points of view of those who form opinion and impress upon us their take on the world.

TOWARD WHOLENESS

At certain points in our history we were better able to synthesize these differing kinds of attention than we are at present. While it is not possible to delve into the sort of cultural analysis which forms the second part of my book here, I want to mention one intriguing indicator for its objective measurability. Early representations of the human face tend to stare straight ahead, and are quite inexpressive. And then, around the sixth century BCE something miraculous happens. We seem to be able to understand faces, and to construct wonderful representations of them in space. At this time, the direction of the gaze in such faces tends to shift toward the viewer's left, the visual field that engages the right hemisphere.[31] The subjects also tend to expose their left cheek, the more expressive part of the face, the aspect controlled by the right hemisphere. Of course these artists knew nothing about brain structure or function; they just intuitively represented the face in this way.

A similar progression happened in Rome—during the Republic and early Augustan era, highly expressive portraiture was supported by a sophisticated understanding of facial musculature. As Rome grew into an empire interested in might, manipulation, and hierarchy, portraiture lost its personal and empathic vision. The Renaissance was another era in which the left and right hemispheres appear to work in harmony with one another; but the shift of gaze in portraiture to the viewer's left that occurred in the fourteenth and fifteenth centuries had faded out again by the twentieth century.[32]

Once you come forward to the Enlightenment, the period when Descartes so famously separated the head from the body, reason was singled out as the most prized human attribute, and it was not until the age of romanticism that a balance between ways of knowing was once again sought. The term romanticism is rather unfortunate, as it makes it sound like a time-limited, culture-bound and self-indulgent phenomenon, when in fact it was during this period that Western philosophy reached one of its peaks. The arts, too, during this time reflect a more complete view of the world, one recognizable to any other society than our contemporary society in the West. It is our current technical, disembodied, and dehumanized view of the world that is foreign to all other cultures and eras of human history.

I would suggest that we now live in a world in which the left-hemisphere view has come to dominate. If I am right, you would see something like this coming about: first and foremost, there would be a loss of the bigger picture. Knowledge would be replaced by tokens, and wisdom would be hard to find. There would be a loss of appreciation of skill and judgment, and both would be replaced by the type of algorithms used by computers. Hubert Dreyfus wrote *What Computers Can't Do* in 1972, and then twenty years later he wrote another book, *What Computers Still Can't Do*.[33] In these works he points out that above a certain level, skill simply cannot be operationalized. You can get a long way and do very well with operationalizable skills, but the genuine expert does not rely on them.

In the left-brain world, bureaucracy would have a field day. Our sense of uniqueness would be lost, quantity would become more important than quality, and our thinking would reflect the "either/or" mentality of the left hemisphere, in its constant battle for (an illusory) clarity. This quest for certainty would drive out an understanding of the conjunction of opposites—one of the greatest insights of the pre-Socratic philosophers, and of philosophers in other cultures, but one that has been lost in the world in which we now live.

Reasonableness would be replaced by rationality, a strict adherence to rules rather than the ability to temper rules with intuition. Intuition can lead you astray, sure, but so can mindless rationalizing. You need to be able both to intuit and to reason. Intuition is not a mere indulgence: it is the precious distillate of experience. People who reason well have better intuitions than people who do not; equally, those with good intuitions reason better than those without them.

Systems would become designed simply to maximize utility. Since the left hemisphere is interested in control, you would tend to see a lack of trust and a sense of paranoia, imaged in the overuse of closed-circuit cameras and genetic data banks. There is a misguided notion that the left hemisphere is cool and objective, but it is not at all. Not only is its "quick and dirty" take on the world inaccurate, it also becomes enraged when challenged. Of all the emotions, anger is the one most clearly lateralized, and it lateralizes to the left.[34] The left hemisphere tends to identify itself as a passive victim of the errors of others—and we would too.

Art would become merely conceptual. Visual art would lack a sense of depth, and would favor distorted or bizarre perspectives. Music would be reduced to little more than rhythm—the only part of music that the left hemisphere, in normal subjects, is capable of processing; language would become diffuse, excessive, and lacking in concrete referents. There would be a deliberate undercutting of the sense of wonder. Awe would be dismissed as mystification.

Flow would be broken down to a series of pieces—effectively digitalized. The tacit forms of knowing on which society utterly depends would be abandoned in favor of "a network of small complicated rules," a phrase de Tocqueville applied when visiting America in the 1830s.[35] We would be spectators rather than actors in the world—an ambition to which Descartes proudly aspired.[36]

And there would be a dangerously unwarranted optimism, because the hemispheres also differ in their degree of optimism. This was first discovered years ago by giving each isolated hemisphere a personality inventory to fill out. This inventory was also given to friends and relatives. It turned out that the left hemisphere had an exaggerated sense of its own talents and virtues, while the right hemisphere, although more realistic, tended to underplay them; as a result the left hemisphere was dubbed "the polisher," and the right "the tarnisher."[37]

Twentieth-century art, with many very noble exceptions, has tended to trivialize what art does and turn it into a clever intellectual game, rather than something that deeply engages us and moves us to see things that we would not otherwise see. Juhani Pallasmaa, Sarah Robinson, and I share many common themes which seem to relate to this difference between the hemispheres. For one thing, architecture has become too academic. When you think of many of the great architects of the past, such as Christopher Wren and Nicholas Hawksmoor, they were primarily stonemasons. Architectural schools as we know them did not yet exist. These were stonemasons who knew their craft and their

5.3 John Singer Sargent, *Pavement of St. Mark's, Venice*, 1880–1882. Oil on canvas.

materials, possessed an intuitive sense of proportion, and knew the ancients. They read the works of Vitruvius, but their emphasis was not on abstract theory; their approach was embodied.

Of course, the spirit of Merleau-Ponty always hangs over all such discussions. I draw on his philosophy in *The Master and His Emissary* and am a great admirer of his work. He emphasized the importance of the senses in constituting both ourselves and our world. But in fact he was doing—explicitly, and for philosophy—what the arts had been doing implicitly for thousands of years. As Pallasmaa says, "Significant architecture makes us experience ourselves as complete, embodied spiritual beings. In fact, this is the great function of all meaningful art."[38]

In the medieval world, even the pavements look good enough to eat. When I first went to Italy, I remember looking at the texture of the surfaces of these foot-polished marble stones and thinking it looked like some wonderful dessert. This is a world that speaks to us through multiple senses, through touch and taste as well as through sight and sound. Remember Goethe in Rome with his loved one, lying in bed and running his hands over the contours of her body, and saying: "Am I not instructing myself by observing the form of her lovely bosom, guiding my hand down over her hips? Then at last I truly understand the marbles: I think and compare, see with a feeling eye, feel with a seeing hand."[39] Heidegger said that a building is often better disclosed in its *essent* (core essence) by a smell than through other modalities.[40]

That wonderful Norwegian writer on late Roman architecture Hans Peter L'Orange showed that the sense of proportion, and the delicacy and modesty of design, that the Romans inherited from the Greeks was lost as the Empire came on and as we moved into a more left-hemisphere-dominated world: the wonderfully proportioned details of old temples were just ripped out and shoved into walls, as so-called *spolia*, makeweights to help fill out the massive concrete walls that empire demanded.[41] In other words, the sense of proportion, which is built on the body, the sense of how the parts relate to one another, the right hemisphere's take—disappeared.

The hand is not just a grasper, it is also creator, a mediator—the hand of the potter, the mason. The hands-on crafts need to be encouraged. We have neglected them, and as a result architecture has become too much of an intellectualized, theoretical business.

Michelangelo depicts the hand of God in the act of creation; there are only two parts of God's anatomy that are ever discussed—the eye of God and the hand of God. And of course the eye can be seen as tyrannical, as Pallasmaa rightly points out. Herder, the German philosopher of early romanticism, said that "a thousand viewpoints are not sufficient" to prevent the living form from being reduced by sight, when unaided by the other senses, to a two-dimensional diagram—what he calls a "pitiful polygon."[42] This fate is avoided only when the viewer's "eye becomes his hand." Herder points to the importance of an unbroken continuity which dismisses as inadequate any mere focus on parts; a never-resting evolution that defies stasis; an insistence on depth, volume, as opposed to the flatness of a single plane of vision; a commitment to the work of art imaged in the urgent recruitment of empathy mediated by the hand, rather than the detached coolness of the eye.

But the eye does not have to be like that. Just as the hand and the eye of God can be used for control, they can also be directed toward empathic connection. It is instructive to consider what happens to the world when sight is removed. There is a wonderful exploration of this in a book called *Touching the Rock* by John Hull, an English academic who gradually went blind in his middle years.[43] He knew what it was to be sighted, and he also knew what happened to his world as he moved into the realm of blindness. He observed that the sense of vision gives you the whole picture, which lends coherence to a world in which sound comes and goes unpredictably. But it also makes the world appear more static, more structured, whereas, without sight, the world appears more in flux, changing and becoming, and then receding. Sight can also connect you with the world, but in a very special sense: in a sense of wanting to get things from it—in other words, as a resource. Sight connects us to what we want; it puts us in an affectively charged relationship with it, in the sense that something must yield itself to us, a relationship that is, in this respect, essentially a possessive, inquisitive one. But vision can also sunder us from the world, and when it acts in this way, it tends to remove us from our body and limit us in space, as if the price of freeing us from physical existence is that we are reduced to a point and our connections are severed.

For the blind person, though, the self is still in a sense confined to the body. That body is something much richer; it expands more easily into the experienced world around him. Blind people live more in and through their body, according to Hull. He says that sound is experienced *within* us; things seen appear to be somewhere *without* us. This is the essence of sight—an objectivity which relates to being in control. You can close your eyes; you can't close your ears.

"You force time to your will," Hull writes about the rest of us. "Time, for sighted people, is that against which they fight. For blind people, time, against which you previously fought, becomes simply a stream of consciousness within which you act. It's as a world that comes to me, which springs into life for me, which has no existence apart from its life towards me and with which I am no longer in conflict to the same degree."[44] Time loses its pressure because it is not measured against desire. Hull is no longer running, but feels himself to be in the flow of time. In other words, he goes more into the right-hemisphere mode of continuity, connection, and flow. But he feels less determined, he feels less of a controlling consciousness. The eye draws attention to itself. It makes us think that our self is behind our eyes, and it obtrudes on the other senses. So, interestingly, blind-sight—echolocation, as he learned to call it—is afforded by the ear, but it is as if the ear takes no credit for it, and does not localize our consciousness to itself. The experience of echolocation is felt, in his amazing description of it, as "distributed throughout the person, in the skin, in the face, in the body. One is not aware of listening. One is simply aware of becoming aware. The sense of pressure is on the skin of the face, rather than upon or within the ears."[45]

Hull's insight into his experience of a sightless world emphasizes how much seeing contributes to the sense of the world as clear, stable, under our control, subject to an active will; and to the sense we have of ourselves standing over against it. And in sharing his insight he indicates the degree to which sight is particularly congenial to the mode of being of the left hemisphere.

But it can also be congenial to the right hemisphere, and for that to happen we need to disrupt focused attention. In other words, we need to pay attention to the whole, and to peripheral vision, in architecture. The qualities of the right hemisphere need to be engaged: indirectness and implicitness, not the overwhelmingly explicit and confrontational; the embodied and sensuous, not the purely cerebral and clinical; flow, harmony, and depth, not disruption, discord, and surfaces that repel the mind, if not the eye; and an ability to evolve and change, without change being considered simply a decline from sterile perfection. Gaze needs to pass through the work of art or architecture, through its surfaces, not arrested at them and rebuffed, in order, as Merleau-Ponty so wonderfully put it, not so much to see the *work*, as to see the *world* "according to it."[46] There is a semitransparency that we miss in a magnificently polished appearance, whereas it was present in the porous surfaces of old stone. We need to engage a spirit of modesty rather than one of grandiosity.

Above all this approach involves a kind of "betweenness,"[47] demonstrated in this illustration by Fra Angelico of *The Annunciation* (figure 5.4). We see here a patient, calm attention and contemplation, rather than experience a sudden shock of excitement. Attention is a kind of love. The French philosopher Louis Lavelle said that love is pure attention to the existence of the other.[48] Often, in our very fast-moving world, we get only an impression, we get a shock image. We really need to live with a building and feel it for a period of time before we understand it, and that experience is not going to be delivered by the eye alone. In *The Annunciation* you can see that the eyes are engaged but also modestly averted in this encounter. The work interacts with us and draws something out of us, and we interact with the work of art and draw something out of it—neither one is ever the same again.

When I first heard Pallasmaa say that architecture is the art of "petrified silence," I immediately recalled Goethe's famous remark that architecture is frozen music. These two things come together because music is about betweenness; it is all about gaps, silences. A great piece like the Schubert C Major Quintet can change the course of your life, but it is compounded only out of notes and the gaps between notes. Each of those sounds on its own is absolutely nothing, is without meaning. Yet music is constructed just by putting individual notes together. So where is it, where is the music? It can only be in the gaps. The gaps make the melody, the gaps make the harmony and rhythm. But the

5.4 Fra Angelico, *The Annunciation*, 1437–1446. Fresco, the Convent of St. Mark's, Florence, Italy.

gaps are just silence; nothing can come out of silence. So where does this thing come from? It comes out of the "betweenness," out of the coming together of these things. And in this respect architecture is like music.

When I entered the pavilion at Taliesin West, I noticed Lao Tzu's saying that the reality of the building is the space inside to be lived in. I would like to tweak the translation: the reality of the building is the space inside to be "lived," rather than "lived *in*." Standing in a great Greek temple, in a Renaissance church by one of the masters, one feels a depth, an unobtrusive simplicity, a sense of spaces calling out and answering one another. I remember as a teenager in Florence going into the church of Santo Spirito and being unable to leave it: every day I had to return. Yet photograph the space and you can hardly see what is so special about it. It looks austere, even simple. But there is something about the feeling of the body in that space that expands the soul, the spirit, the heart, the mind, and one's physical senses. This is, I think, the signature of a great work of art. Interestingly, after Brunelleschi left the building of San Lorenzo to work elsewhere, its ground plan, which had been almost identical to that of Santo Spirito, was subtly altered. The original plan was not followed precisely, so the proportions changed: perhaps that is why I have never found San Lorenzo in any way as captivating as Santo Spirito. Later in my life, almost in spite of my better judgment, I came face to face with Balthasar Neumann's late baroque masterpiece, the Vierzehnheiligen in Bavaria. There is no way you can capture what this building feels like from the eye. You need to be in it. I was prepared to be enraptured by all sorts of details, because it is a singularly ornate church, but in fact

5.5 Balthasar Neumann, Vierzehnheiligen, near Staffelstein, Germany, 1743–1772.

the most surprising thing about it is the calm sense of unified space within it, which exerts a magnetic attraction. I tried unsuccessfully three times to leave the church before finally tearing myself away.

We instinctively feel repulsed by inhumanly smooth surfaces, rebuffed by the flatness that is present in so much architecture of the last few decades, which may have had a grand design, but is not configured to the human form. Natural materials allow us to sink in; they have a history—a history that is revealed in the coming together of space and time in the present. Weathering, transition, decay: we need to permit them, even embrace them. They are a part of a changing world, not a perfect, static world. Unfortunately, machines, like my Apple laptop, only get to look scratched and dirty when time marks them. They do not get more beautiful, like the old writing desk in my study. There is a relational, dynamic interactive understanding of space and time that we need to accept. An enclosed space can open and expand the physical sense of one's self, not rebuff it. An enclosed time can open to eternity and expand the sense of the moment. You do not get to eternity by turning your back on time, but by going through the region of the temporal. You do not get to infinity by turning your back on the finite, but on the contrary by embracing the finite, going into and passing through it—to emerge on the other side.

5.6 Frank Lloyd Wright, Fallingwater, Mill Run, Pennsylvania, 1935–1937: hearth.

We live in a world of false perfection. Architecture has become an extension of our egos, an intrusion into the natural world rather than an extension of nature into the man-made realm. How important it is to feel we belong here! I was delighted to discover, in writing *The Master and His Emissary*, that many animals bond as much with their nests as they do with their mothers. Think of that. The mother is the most important thing in the world to any creature, and home is right up there with motherhood. Can we no longer create places where we feel we belong?

The Latin word *focus* means the hearth. What is so marvelous about the hearth is that, when we sit around it, we cease to focus on one another, because we are all focusing, together, on the fire. That allows indirect attention to one's companions, an open, undemanding, friendly rather than inquisitorial, attention: whereas sitting in a fireless room in an armchair eye-to-eye, we feel at once on guard. That warmth, and the indirect gaze that is implicit, are important. We hunger for old, grained surfaces, made of materials that can show their age with grace, and for areas where things are half-seen, rather than exposed to the full glare of attention.

So although it may be important to strive to understand the brain, it is even more important to understand that the brain is embodied, and embedded in culture, as are we. What you know through your experience as architects cannot be improved on by learning about the brain function associated with it, any more than your understanding of a documentary on television depends on knowing about the workings of the set. When I go and talk to dancers, for example, they want to know what is going on in their brains while they dance. I fear this is not likely to help, because they are the experts on what they are doing when they dance, not I. You cannot get beyond experience—and in experience we are all, literally, the "experts," for that is what the word means. Remember that your understanding comes through your body. You experience the world intuitively through your senses more than you can ever know the world intellectually. Memory, the traces that experience leaves in us like a magnetic imprint, is stored in the gut, muscles, and bone as well as the brain. By all means learn about neuroscience—but the best lesson to learn from it is that only one half of your brain, the half that sees and understands the least, would ever think that great architecture can be reduced to the brain.

NOTES

1. I. McGilchrist, *The Master and His Emissary: The Divided Brain and the Making of the Western World* (New Haven: Yale University Press, 2009).

2. L. Jäncke and H. Steinmetz, "Anatomical Brain Asymmetries and Their Relevance for Functional Asymmetries," in K. Hugdahl and R. J. Davidson, eds., *The Asymmetrical Brain* (Cambridge, MA: MIT Press, 2003), 187–230, 210–211.

3. F. Conti and T. Manzoni, "The Neurotransmitters and Postsynaptic Actions of Callosally Projecting Neurons," *Behavioral Brain Research* 64 (1994): 37–53; B.-U. Meyer, S. Röricht, H. Gräfin von Einsiedel, et al., "Inhibitory and Excitatory Interhemispheric Transfers between Motor Cortical Areas in Normal Subjects and Patients with Abnormalities of the Corpus Callosum," *Brain* 118 (1995): 429–440; S. Röricht, K. Irlbacher, E. Petrow, et al., "Normwerte transkallosal und kortikospinal vermittelter Effekte einer hemisphärenselektiven elektromyographischer magnetischen Kortexreizung beim Menschen," *Zeitschrift für Elektroenzephalographie, Elektromyographie und Verwandte Gebiete* 28 (1997): 34–38; J. Höppner, E. Kunesch, J. Buchmann, et al., "Demyelination and Axonal Degeneration in Corpus Callosum Assessed by Analysis of Transcallosally Mediated Inhibition in Multiple Sclerosis," *Clinical Neurophysiology* 110 (1999): 748–756; C. D. Saron, J. J. Foxe, G. V. Simpson, et al., "Interhemispheric Visuomotor Activation: Spatiotemporal Electrophysiology Related to Reaction Time," in E. Zaidel and M. Iacoboni, eds., *The Parallel Brain: The Cognitive Neuroscience of the Corpus Callosum* (Cambridge, MA: MIT Press 2002), 171–219.

4. J. Thurman, "On the Weight of the Brain and the Circumstances Affecting It," *Journal of Mental Science* 12 (1866): 1–43; J. Crichton-Browne, "On the Weight of the Brain and Its Component Parts in the Insane," *Brain* 2 (1880): 42–67; G. von Bonin, "Anatomical Asymmetries of the Cerebral Hemispheres," in V. B. Mountcastle, ed., *Interhemispheric Relations and Cerebral Dominance* (Baltimore: Johns Hopkins University Press, 1962), 1–6; H. Hadziselimovic and H. Cus, "The Appearance of Internal Structures of the Brain in Relation to Configuration of the Human Skull," *Acta Anatomica* 63 (1966): 289–299; A. M. Galaburda, M. LeMay, T. L. Kemper, et al., "Right-Left Asymmetrics in the Brain," *Science* 199 (1978): 852–856; M. LeMay, "Morphological Aspects of Human Brain Asymmetry: An Evolutionary Perspective," *Trends in Neurosciences* 5 (1982): 273–275; M. Schwartz, H. Creasey, C. L. Grady, et al., "Computed Tomographic Analysis of Brain Morphometrics in 30 Healthy Men, Aged 21 to 81 Years," *Annals of Neurology* 17 (1985): 146–157; S. Weis, H. Haug, B. Holoubek, et al., "The Cerebral Dominances: Quantitative Morphology of the Human Cerebral Cortex," *International Journal of Neuroscience* 47 (1989): 165–168; A. Kertesz, M. Polk, S. E. Black, et al., "Anatomical Asymmetries and Functional Laterality," *Brain* 115 (1992): 589–605; K. Zilles, A. Dabringhaus, S. Geyer, et al., "Structural Asymmetries in the Human Forebrain and the Fore-Brain of Non-Human Primates and Rats," *Neuroscience and Biobehavioral Reviews* 20 (1996): 593–605; J. N. Zilles, J. W. Snell, N. Lange, et al., "Quantitative Magnetic Resonance Imaging of Human Brain Development: Ages 4–18," *Cerebral Cortex* 6 (1996): 551–560; H. Damasio, *Human Brain Anatomy in Computerized Images*, 2nd edn. (Oxford: Oxford University Press, 2005).

5. A. M. Galaburda, "Anatomic Basis of Cerebral Dominance," in R. J. Davidson and K. Hugdahl, eds., *Brain Asymmetry* (Cambridge, MA: MIT Press, 1995), 51–73.

6. R. C. Gur, I. K. Packer, J. P. Hungerbühler, et al., "Differences in the Distribution of Gray and White Matter in Human Cerebral Hemispheres," *Science* 207 (1980): 1226–1228; Galaburda, "Anatomic Basis of Cerebral Dominance"; R. C. Gur, B. I. Turetsky, M. Matsui, et al., "Sex Differences in Brain Gray and White Matter in Healthy Young Adults: Correlations with Cognitive Performance," *Journal of Neuroscience* 19 (1999): 4065–4072; J. S. Allen, H. Damasio, T. J. Grabowski, et al., "Sexual Dimorphism and Asymmetries in the Gray-White Composition of the Human Cerebrum," *NeuroImage* 18 (2003): 880–894.

7. T. L. Hayes and D. A. Lewis, "Hemispheric Differences in Layer III Pyramidal Neurons of the Anterior Language Area," *Archives of Neurology* 50 (1993): 501–505.

8. D. Lewis and M. C. Diamond, "The Influence of Gonadal Steroids on the Asymmetry of the Cerebral Cortex," in Davidson and Hugdahl, *Brain Asymmetry*, 31–50.

9. S. D. Glick, D. A. Ross, and L. B. Hough, "Lateral Asymmetry of Neurotransmitters in Human Brain," *Brain Research* 234 (1982): 53–63; H. N. Wagner, Jr., H. D. Burns, R. F. Dannals, et al., "Imaging Dopamine Receptors in the Human Brain by Positron Emission Tomography," *Science* 221 (1983): 1264–1266; D. M. Tucker and P. A. Williamson, "Asymmetric Neural Control Systems in Human Self-Regulation [review]," *Psychological Review* 91 (1984): 185–215.

10. See, e.g., L. J. Rogers and R. J. Andrew, *Comparative Vertebrate Lateralization* (Cambridge: Cambridge University Press, 2002), for an overview.

11. O. Güntürkün, B. Diekamp, M. Manns, et al., "Asymmetry Pays: Visual Lateralization Improves Discrimination Success in Pigeons," *Current Biology* 10 (2000): 1079–1081; L. J. Rogers, P. Zucca, and G. Vallortigara, "Advantages of Having a Lateralized Brain," *Proceedings of the Royal Society of London, Series B: Biological Sciences* 271(suppl. 6) (2004): S420–422.

12. See McGilchrist, *The Master and His Emissary*, 25ff.

13. A. H. van Zomeren and W. H. Brouwer, *Clinical Neuropsychology of Attention* (Oxford: Oxford University Press, 1994); and see McGilchrist, *The Master and His Emissary*, 38–40, for full discussion.

14. M. Leclercq, "Theoretical Aspects of the Main Components and Functions of Attention," in M. Leclercq and P. Zimmerman, eds., *Applied Neuropsychology of Attention* (London: Psychology Press, 2002), 3–55, 16.

15. D. Navon, "Forest before Trees: The Precedence of Global Features in Visual Perception," *Cognitive Psychology* 9 (1977): 353–383; D. E. Broadbent, "The Hidden Preattentive Process," *American Psychologist* 32 (1977): 109–118; R. D. Nebes, "Direct Examination of Cognitive Function in the Right and Left Hemispheres," in M. Kinsbourne, ed., *Asymmetrical Function of the Brain* (Cambridge: Cambridge University Press, 1978), 99–137; A. Young and G. Ratcliff, "Visuospatial Abilities of the Right Hemisphere," in A. W. Young, ed., *Functions of the Right Cerebral Hemisphere* (London: Academic Press, 1983), 1–31; E. Zaidel, "Language in the Right Hemisphere," in D. Benson and E. Zaidel, eds., *The Dual Brain: Hemispheric Specialization in Humans* (New York: Guilford Press, 1985), 205–231; J. B. Hellige, "Hemispheric Asymmetry for Components of Visual Information Processing," in Davidson and Hugdahl, *Brain Asymmetry*, 99–121; S. Christman, *Cerebral Asymmetries in Sensory and Perceptual Processing* (Amsterdam: Elsevier, 1997); M. J. Beeman, E. M. Bowden, and M. A. Gernsbacher, "Right and Left Hemisphere Cooperation for Drawing Predictive and Coherence Inferences during Normal Story Comprehension," *Brain and Language* 71 (2000): 310–336.

16. Navon, "Forest before Trees"; G. R. Mangun, S. J. Luck, R. Plager, et al., "Monitoring the Visual World: Hemispheric Asymmetries and Subcortical Processes in Attention," *Journal of Cognitive Neuroscience* 6 (1994): 267–275.

17. J. Dewey, *Context and Thought*, University of California Publications in Philosophy 12, no. 3 (Berkeley: University of California Press, 1931).

18. M. C. Corballis, "Hemispheric Interactions in Temporal Judgments about Spatially Separated Stimuli," *Neuropsychology* 10 (1996): 42–50; M. C. Corballis, L. Boyd, A. Schulze, et al., "Role of the Commissures in Interhemispheric Temporal Judgments," *Neuropsychology* 12 (1998): 519–525.

19. R. Dunbar, *The Human Story: A New History of Mankind's Evolution* (London: Faber, 2004); M. Durnford and D. Kimura, "Right Hemisphere Specialization for Depth Perception Reflected in Visual Field Differences," *Nature* 231 (1971): 394–395; M. S. Gazzaniga and J. E. LeDoux, *The Integrated Mind* (New York: Plenum Press, 1978); A. Y. Egorov and N. N. Nikolaenko, "Functional Brain Asymmetry and Visuo-Spatial Perception in Mania, Depression and Psychotropic Medication," *Biological Psychiatry* 32 (1992): 399–410; N. N. Nikolaenko, A. Y. Egorov, and E. A. Freiman, "Representation Activity of the Right and Left Hemispheres of the Brain," *Behavioral Neurology* 10 (1997): 49–59; N. N. Nikolaenko and A. Y. Egorov, "The Role of the Right and Left Cerebral Hemispheres in Depth Perception," *Fiziologiia Cheloveka* (Human Physiology) 24, no. 6 (1998): 21–31; N. N. Nikolaenko, "Representation Activity of the Right and Left Hemispheres of the Brain," *Acta Neuropsychologica* 1 (2003): 34–47.

20. H.-G. Wieser and G. Mazzola, "Musical Consonances and Dissonances: Are They Distinguished Independently by the Right and Left Hippocampi?," *Neuropsychologia* 24 (1986): 805–812; A. Preisler, E. Gallasch, and G. Schulter, "Hemispheric Asymmetry and the Processing of Harmonies in Music," *International Journal of Neuroscience* 47 (1989), 131–140; M. J. Tramo and J. J. Bharucha, "Musical Priming by the Right Hemisphere Post-callosotomy," *Neuropsychologia* 29 (1991): 313–325; S. Evers, J. Dannert, D. Rodding, et al., "The Cerebral Hemodynamics of Music Perception: A Transcranial Doppler Sonography Study," *Brain* 122 (1999): 75–85; N. Passynkova, H. Neubauer, and H. Scheich, "Spatial Organization of EEG Coherence during Listening to Consonant and Dissonant Chords," *Neuroscience Letters* 412 (2007): 6–11.

21. D. Kimura, "Right Temporal-Lobe Damage: Perception of Unfamiliar Stimuli after Damage," *Archives of Neurology* 8 (1963): 264–271; H. Gardner, *The Shattered Mind* (New York: Knopf, 1974);

T. G. Bever and R. J. Chiarello, "Cerebral Dominance in Musicians and Nonmusicians," *Science* 185 (1974): 537–539; H. W. Gordon and A. Carmon, "Transfer of Dominance in Speed of Verbal Response to Visually Presented Stimuli from Right to Left Hemisphere," *Perceptual and Motor Skills* 42 (1976): 1091–1100; B. Cotton, O. J. Tzeng, and C. Hardyck, "Role of Cerebral Hemispheric Processing in the Visual Half-Field Stimulus-Response Compatibility Effect," *Journal of Experimental Psychology: Human Perception and Performance* 6 (1980): 13–23; R. W. Sperry, "Consciousness, Personal Identity and the Divided Brain," in Benson and Zaidel, *The Dual Brain*, 11–26; M. Regard and T. Landis, "Beauty May Differ in Each Half of the Eye of the Beholder," in I. Rentschler, B. Herzberger, and D. Epstein, eds., *Beauty and the Brain: Biological Aspects of Aesthetics* (Basel: Birkhäuser, 1988), 243–256; D. J. Thal, V. Marchman, J. Stiles, et al., "Early Lexical Development in Children with Focal Brain Injury," *Brain and Language* 40 (1991): 491–527; R. J. Haier, B. V. Siegel, A. MacLachlan, et al., "Regional Glucose Metabolism Changes after Learning a Complex Visuospatial/Motor Task: A Positron Emission Tomographic Study," *Brain Research* 570 (1992): 134–143; D. L. Mills, S. A. Coffey-Corina, and H. J. Neville, "Language Acquisition and Cerebral Specialization in 20-Month-Old Infants," *Journal of Cognitive Neuroscience* 5 (1993): 317–334; M. E. Raichle, J. A. Fiez, T. O. Videen, et al., "Practice-Related Changes in Human Brain Functional Anatomy during Nonmotor Learning," *Cerebral Cortex* 4 (1994): 8–26; J. M. Gold, K. F. Berman, C. Randolph, et al., "PET Validation of a Novel Prefrontal Task: Delayed Response Alternation (DRA)," *Neuropsychology* 10 (1996): 3–10; E. Tulving, H. J. Markowitsch, F. E. Craik, et al., "Novelty and Familiarity Activations in PET Studies of Memory Encoding and Retrieval," *Cerebral Cortex* 6 (1996): 71–79; R. Shadmehr and H. H. Holcomb, "Neural Correlates of Motor Memory Consolidation," *Science* 277 (1997): 821–825; G. S. Berns, J. D. Cohen, and M. A. Mintun, "Brain Regions Responsive to Novelty in the Absence of Awareness," *Science* 276 (1997): 1272–1275; J. Cutting, *Principles of Psychopathology* (Oxford: Oxford University Press, 1997), 67; A. Martin, C. L. Wiggs, and J. Weisberg, "Modulation of Human Medial Temporal Lobe Activity by Form, Meaning and Experience," *Hippocampus* 7 (1997): 587–593; L. J. Rogers, "Evolution of Hemisphere Specialization: Advantages and Disadvantages," *Brain and Language* 73 (2000): 236–253; M. A. Persinger and C. A. Lalonde, "Right to Left Hemispheric Shift in Occipital Electroencephalographic Responses to Repeated Kimura Figures," *Perceptual and Motor Skills* 91 (2000): 273–278; R. Henson, T. Shallice, and R. Dolan, "Neuroimaging Evidence for Dissociable Forms of Repetition Priming," *Science* 287 (2000): 1269–1272; J. S. Feinstein, P. R. Goldin, M. B. Stein, et al., "Habituation of Attentional Networks during Emotional Processing," *NeuroReport* 13 (2002): 1255–1258; V. Treyer, A. Buck, and A. Schnider, "Subcortical Loop Activation during Selection of Currently Relevant Memories," *Journal of Cognitive Neuroscience* 15 (2003): 610–618.

22. E. Goldberg and L. D. Costa, "Hemispheric Differences in the Acquisition and Use of Descriptive Systems," *Brain and Language* 14 (1981): 144–173; E. Goldberg, "Associative Agnosias and the Functions of the Left Hemisphere," *Journal of Clinical and Experimental Neuropsychology* 12 (1990): 467–484; E. Goldberg, K. Podell, and M. Lovell, "Lateralization of Frontal Lobe Functions and Cognitive Novelty," *Journal of Neuropsychiatry and Clinical Neurosciences* 6 (1994): 371–378; E. Goldberg, *The Executive Brain: Frontal Lobes and the Civilized Mind* (Oxford: Oxford University Press, 2001).

23. See McGilchrist, *The Master and His Emissary*, 79–83.

24. J. Ruskin, *Modern Painters*, 6 vols. (London: George Allen, 1904), vol. 4, 60–61.

25. H. Gardner, P. K. Ling, L. Flamm, et al., "Comprehension and Appreciation of Humorous Material Following Brain Damage," *Brain* 98 (1975): 399–412; E. Winner and H. Gardner, "The Comprehension of Metaphor in Brain-Damaged Patients," *Brain* 100 (1977): 717–729; W. Wapner, S. Hamby, and H. Gardner, "The Role of the Right Hemisphere in the Apprehension of Complex Linguistic Materials," *Brain and Language* 14 (1981): 15–33; H. H. Brownell, D. Michel, J. Powelson, et al., "Surprise but Not Coherence: Sensitivity to Verbal Humor in Right-Hemisphere Patients," *Brain and Language* 18 (1983): 20–27; M. Dagge and W. Hartje, "Influence of Contextual Complexity on the Processing of

Cartoons by Patients with Unilateral Lesions," *Cortex* 21 (1985): 607–616; A. M. Bihrle, H. H. Brownell, J. A. Powelson, et al., "Comprehension of Humorous and Nonhumorous Materials by Left and Right Brain-Damaged Patients," *Brain and Cognition* 5 (1986): 399–411; J. A. Kaplan, H. H. Brownell, J. R. Jacobs, et al., "The Effects of Right Hemisphere Damage on the Pragmatic Interpretation of Conversational Remarks," *Brain and Language* 38 (1990): 315–333; H. H. Brownell, T. L. Simpson, A. M. Bihrle, et al., "Appreciation of Metaphoric Alternative Word Meanings by Left and Right Brain-Damaged Patients," *Neuropsychologia* 28 (1990): 375–383; M. Beeman, "Semantic Processing in the Right Hemisphere May Contribute to Drawing Inferences from Discourse," *Brain and Language* 44 (1993): 80–120; D. Anaki, M. Faust, and S. Kravetz, "Cerebral Hemisphere Asymmetries in Processing Lexical Metaphors" (1), *Neuropsychologia* 36 (1998): 353–362; D. Anaki, M. Faust, and S. Kravetz, "Cerebral Hemisphere Asymmetries in Processing Lexical Metaphors" (2), *Neuropsychologia* 36 (1998): 691–700; P. Shammi and D. T. Stuss, "Humor Appreciation: A Role of the Right Frontal Lobe," *Brain* 122 (1999): 657–666.

26. I. McGilchrist, *Against Criticism* (London: Faber, 1982).

27. Winner and Gardner, "The Comprehension of Metaphor in Brain-Damaged Patients"; Brownell, Simpson, Bihrle, et al., "Appreciation of Metaphoric Alternative Word Meanings by Left and Right Brain-Damaged Patients"; Anaki, Faust, and Kravetz, "Cerebral Hemisphere Asymmetries in Processing Lexical Metaphors" (1) and (2).

28. M. B. Bender, M. Feldman, and A. J. Sobin, "Palinopsia," *Brain* 9 (1968): 321–338; E. M. Michel and B. T. Troost, "Palinopsia: Cerebral Localization with CT," *Neurology* 30 (1980): 887–889; T. Müller, T. Büttner, W. Kuhn, et al., "Palinopsia as Sensory Epileptic Phenomenon," *Acta Neurologica Scandinavica* 91 (1995): 433–436; J. L. Cummings, "Neuropsychiatric Manifestations of Right Hemisphere Lesions," *Brain and Language* 57 (1997): 22–37; S. Schwartz, F. Assal, N. Valenza, et al., "Illusory Persistence of Touch after Right Parietal Damage: Neural Correlates of Tactile Awareness," *Brain* 128 (2005): 277–290; M. E. Ritsema and M. A. Murphy, "Palinopsia from Posterior Visual Pathway Lesions without Visual Field Defects," *Journal of Neuro-Ophthalmology* 27 (2007): 115–117.

29. S. Robinson, *Nesting: Body, Dwelling, Mind* (Richmond, CA: William Stout, 2011).

30. V. S. Ramachandran, "Phantom Limbs, Neglect Syndromes, Repressed Memories, and Freudian Psychology," *International Review of Neurobiology* 37 (1994): 291–333.

31. M. Brener, *Faces: The Changing Look of Humankind* (Lanham, MD: University Press of America, 2000); H.-J. Hufschmidt, "Das Rechts-Links-Profil im kulturhistorischen Längsschnitt," *Archiv für Psychiatrie und Nervenkrankheiten* 229 (1980): 17–43; H.-J. Hufschmidt, "Über die Linksorientierung der Zeichnung und die optische Dominanz der rechten Hirnhemisphäre," *Zeitschrift für Kunstgeschichte* 46 (1983): 287–294.

32. I. C. McManus and N. K. Humphrey, "Turning the Left Cheek," *Nature* 243 (1973): 271–272; N. K. Humphrey and I. C. McManus, "Status and the Left Cheek," *New Scientist* 59 (1973): 437–439; O.-J. Grüsser, T. Selke, and B. Zynda, "Cerebral Lateralisation and Some Implications for Art, Aesthetic Perception and Artistic Creativity," in Rentschler, Herzberger, and Epstein, *Beauty and the Brain*, 257–293; R. Latto, "Turning the Other Cheek: Profile Direction in Self-Portraiture," *Empirical Studies of the Arts* 14 (1996): 89–98.

33. H. Dreyfus, *What Computers Can't Do* (New York: HarperCollins, 1972); H. Dreyfus, *What Computers Still Can't Do* (Cambridge, MA: MIT Press, 1992). See also H. Dreyfus and S. Dreyfus, *Mind over Machine: The Power of Human Intuition and Expertise in the Era of the Computer* (New York: Free Press, 1986).

34. E. Harmon-Jones and J. J. Allen, "Anger and Frontal Brain Activity: EEG Asymmetry Consistent with Approach Motivation Despite Negative Affective Valence," *Journal of Personality and Social*

Psychology 74 (1998): 1310–1316; T. Indersmitten and R. C. Gur, "Emotion Processing in Chimeric Faces: Hemispheric Asymmetries in Expression and Recognition of Emotions," *Journal of Neuroscience* 23 (2003): 3820–3825; E. Harmon-Jones, "Contributions from Research on Anger and Cognitive Dissonance to Understanding the Motivational Functions of Asymmetrical Frontal Brain Activity," *Biological Psychology* 67 (2004): 51–76; E. Harmon-Jones, "Trait Anger Predicts Relative Left Frontal Cortical Activation to Anger-Inducing Stimuli," *International Journal of Psychophysiology* 66 (2007): 154–160.

35. A. de Tocqueville, *Democracy in America*, trans. H. Reeve and E. W. Plaag (1840; New York: Barnes and Noble, 2003), 723–724.

36. R. Descartes, "Discourse on the Method of Rightly Conducting the Reason and Seeking for Truth in the Sciences," in *The Philosophical Writings of Descartes*, 3 vols., trans. J. Cottingham, R. Stoothoff, and D. Murdoch (1637; Cambridge: Cambridge University Press, 1984–1991), vol. 1, 109–151, 125.

37. D. M. Bear and P. Fedio, "Quantitative Analysis of Interictal Behavior in Temporal Lobe Epilepsy," *Archives of Neurology* 34 (1977): 454–467.

38. J. Pallasmaa, *The Eyes of the Skin: Architecture and the Senses* (London: Wiley, 2005), 11.

39. J. W. von Goethe, *Römische Elegien*, V, lines 7–10.

40. M. Heidegger, *An Introduction to Metaphysics*, trans. R. Manheim (1935; New Haven: Yale University Press, 1959), 33.

41. H. P. L'Orange, *Art Forms and Civic Life in the Late Roman Empire* (Princeton: Princeton University Press, 1965).

42. J. G. Herder, *Sculpture: Some Observations on Shape and Form from Pygmalion's Creative Dream*, trans. and ed. J. Gaiger (1778; Chicago: University of Chicago Press, 2002), 40–41.

43. J. Hull, *Touching the Rock: An Experience of Blindness* (London: SPCK, 1990).

44. Ibid., 60–61.

45. Ibid., 21.

46. M. Merleau-Ponty, *The Primacy of Perception: And Other Essays on Phenomenology, Psychology, the Philosophy of Art, History and Politics*, ed. J. M. Edie (Evanston: Northwestern University Press, 1964), 164.

47. McGilchrist, *The Master and His Emissary*.

48. L. Lavelle, *L'Erreur de Narcisse* (Paris: Floch, 1939), ch. 9, §7.

6

ARCHITECTURE AND NEUROSCIENCE: A DOUBLE HELIX

John Paul Eberhard

"How would you like to move back to Washington and organize a program to respond to Dr. Salk's challenge?" Norman Koonce, the President of the American Architectural Foundation, and Syl Damianos, Chancellor of the College of Fellows of the American Institute of Architects, asked. At sixty-eight, I had recently retired as Head of the architecture program at Carnegie Mellon University, and was not expecting to return to work again—but how could I refuse such an offer?

Dr. Jonas Salk was convinced that architectural settings profoundly influence our mental and physical welfare, a conviction that stems from his personal experience. While he was working in his laboratory at the University of Pittsburgh School of Medicine, Dr. Salk faced the problem of brain overload. In 1948 he set out to quantify the different types of polio virus, but soon extended his mission to developing a vaccine against polio. For seven years, he and a skilled research team addressed what most people at the time considered the most frightening public health problem in the United States. In 1952 there were nearly 58,000 cases of polio reported; 3,145 people died and 21,269 were left with mild to disabling paralysis; and most of its victims were children.

Dr. Salk realized the importance of his work while he was watching children playing and recognized that thousands of them might never walk again if they contracted polio. He accepted this awesome responsibility, and drove himself at a frantic pace. It was at this

6.1 The basilica of Assisi and the friary Sacro Convento, Assisi, Italy.

point he felt his brain was "overloaded" and that he needed to get away to reinvigorate himself.

Though the motives for his choice of destination are unknown, Dr. Salk decided to go to the basilica at Assisi in Italy for his retreat. The architecture at Assisi is a synthesis of the Romanesque and Gothic styles, creating what was to become known as Italian Gothic architecture. This basilica was designed on two levels, each of which is consecrated as a church. Architecturally, the exterior of the basilica appears united with the friary of St. Francis, since the lofty arcades of the structure support and buttress the church in its apparently precarious position on the hillside. Next to the basilica stands the friary Sacro Convento, whose imposing walls were built with Romanesque arches and powerful buttresses that tower over the valley below, giving the impression of a fortress. The friary now houses a large library and a museum with works of art donated over the centuries by visiting pilgrims. It is likely that Dr. Salk was a guest of the brothers of the friary and was given a room in which to stay during his time there.

His experience at Assisi left such deep impression on him that many years later, Dr. Salk credited the architectural setting there with helping him make the intellectual breakthrough

6.2 Louis Kahn, Salk Institute of Biological Sciences, La Jolla, California, 1959–1965.

that ultimately led to the creation of the polio vaccine. At Assisi he realized that a vaccine based on the inactivated (dead) poliovirus that was injected in humans could be used to combat polio. It was there that he further conceived of producing the vaccine based on three "wild and virulent" reference strains being grown in a type of monkey kidney tissue culture which are then inactivated with formalin.[1] Once he returned to his laboratories in Pittsburgh, the initial trial tests eventually expanded into the largest medical experiment in history, involving 1.8 million children in 44 states. After the Salk vaccine was licensed in 1955 the annual number of polio cases fell from 35,000 in 1953 to 161 cases recorded in 1961. Later a second vaccine, developed by Dr. Albert Sabin, was added for use throughout the world. The result of using the two vaccines has essentially eradicated polio from most countries in the world.

It is interesting that the Salk Institute in La Jolla, California—built in 1963, with Louis Kahn as the architect—has the same relationship to the sea as Assisi has to its surrounding landscape. It also functions very much like a monastery, having quarters that enable the senior scientists who work there to retreat to a place of silence. In 1994, when the Salk Institute received the American Institute of Architects twenty-five-year award, at the ceremony Dr. Salk told this story to the AIA Executive Board and suggested they explore the issue of how architectural settings influence the brain—and consequently behavior. The American Architectural Foundation was given the task of following up on Dr. Salk's challenge, and they recruited me to lead this effort.

I was fascinated with this story, and very interested in moving back to Washington. However, I had to honestly tell Norman and Syl that I had no idea of how to deal intellectually with such a challenge. They replied that neither did they, but they proposed to give me the title of "Director of Discovery" of the AAF. Thus began an intellectual journey that has carried me for nineteen years into the exciting world of neuroscience. In what follows I hope to show how my training in architecture and my scholarly pursuit of neuroscience form a double helix not unlike that of DNA. I am one example of how we use the same brain for each of these activities—with variations in the way neurons and glial cells are organized in networks. Architects' neural networks do not differ from those of neuroscientists, when viewed under an electronic microscope, but the results of their thinking—architectural ideas and neuroscience hypotheses—would be considered by most people to be altogether different. Yet philosopher Patricia Churchland reminds us: "Their brain is what makes humans capable of painting the Sistine Chapel, designing airplanes and transistors, ice skating, reading Shakespeare, and playing Chopin. It is a truly astonishing and magnificent kind of 'wonder-tissue.' Whatever self-esteem justly derives from our accomplishments does so because of the brain, not in spite of it."[2]

THE DOUBLE HELIX OF ARCHITECTURE AND NEUROSCIENCE

Nature provides a relatively simple method for one generation to pass on to the next generation a "blueprint" for each person via the DNA contained in each cell. The structure of DNA—the way it is put together—was discovered by James Watson and Francis Crick in 1953. They disclosed the secret structure of our genes by combining their well-developed scientific knowledge with a leap of the imagination. They determined that the structure is a double-helix polymer, a spiral consisting of two DNA strands wound around each other. These strands of nucleic acids are made up of chains of many repeating units called nucleotides—not unlike the stones in Gothic vaulting. However, stone is a dense, homogeneous material and nucleic acid molecules are incredibly complex, containing the code that guarantees the accurate ordering of the 20 amino acids in proteins made by living cells. Surprisingly, though, there are only a few different nucleotides: there are four different nucleotide units comprising DNA, and only two pairs are possible. In DNA, Adenine (A for short) always pairs with thiamine (T), and guanine (G) always pairs with cytosine (C). It is this feature of complementary base pairing that insures an exact duplicate of each DNA molecule will be passed to its daughter cells when a cell divides.

In much the same way that it has become possible to read the "architectural" design of individuals contained in DNA blueprints, we use our brain structure to calculate proportions in buildings and related objects. This includes how the neurons in our brain represent architectural artifacts like the pyramids as well as the development of neuroscience constructs like DNA. The neurons that accomplish this feat are more or less identical, but they are arranged in billions of networks of untold variety.

THE GOLDEN MEAN

The Greek letter phi represents the "golden ratio," also called the golden section or golden mean. Many architects and artists have proportioned their works to approximate the golden ratio—especially in the form of the golden rectangle, in which the ratio of the longest side to the shorter is the golden ratio—believing this proportion to be aesthetically pleasing. The Egyptian pyramids have sometimes been called golden pyramids because the slant height (along the bisector of the base) is equal to π times the semi-base (half the base width). The isosceles triangle that is the face of such a pyramid can be constructed from the two halves of a diagonally split golden rectangle. I speculate that neuronal circuits in our brain are wired with the natural ability to recognize the golden ratio while we are still embryos—just as we seem to be programmed before birth with a common image of a house that children throughout the world consistently draw when they are five years old.

Vitruvius, the ancient Roman architect, carried the idea of proportions still further: the Vitruvian Man is a drawing created by Leonardo da Vinci circa 1490, that is accompanied by notes based on the work of Vitruvius. The drawing depicts a male figure in two superimposed positions with his arms and legs apart and simultaneously inscribed in a circle and square. The drawing and text are sometimes called the "Canon of Proportions." The drawing is based on the correlations of ideal human proportions with geometry described by Vitruvius in Book III of his treatise *De architectura*. Vitruvius described the human figure as the principal source of proportion among the classical orders of architecture. He determined that the ideal body should be eight heads high. Leonardo's drawing is traditionally named in honor of the architect.

For the human body is so designed by nature that the face, from the chin to the top of the forehead and the lowest roots of the hair, is a tenth part of the whole height; the open

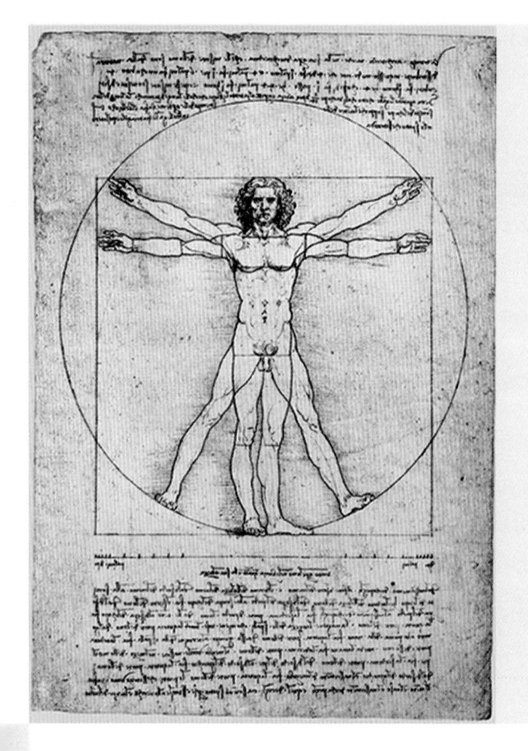

FIGURE 6.3 Leonardo da Vinci, Vitruvian Man, Gallerie dell'Accademia, Venice, 1490.

hand from the wrist to the tip of the middle finger is just the same; the head from the chin to the crown is an eighth, and with the neck and shoulder from the top of the breast to the lowest roots of the hair is a sixth; from the middle of the breast to the summit of the crown is a fourth. If we take the height of the face itself, the distance from the bottom of the chin to the underside of the nostrils is one third of it; the nose from the underside of the nostrils to a line between the eyebrows is the same; from there to the lowest roots of the hair is also a third, comprising the forehead. The length of the foot is one sixth of the height of the body; of the forearm, one fourth; and the breadth of the breast is also one fourth. The other members, too, have their own symmetrical proportions, and it was by employing them that the famous painters and sculptors of antiquity attained to great and endless renown.[3]

Recognizing the proportions of a golden rectangle utilizes neuronal circuits laid down in the frontal cortex and areas of the cingulated gyros. These same circuits, or closely related ones, are used to imagine the double helix of DNA. Thus the architectural construct of a golden mean is processed in the human brain in much the same way, and in more or less the same area of the brain, as the neuroscience construct of a double helix is processed. They both require a brain that can imagine elements in three dimensions without seeing them, and forming images that can be learned, remembered, and taught to novices. The rules for proportion or DNA are *discovered* rather than consciously constructed and then applied.

THE PERCEPTUAL HOUSE OF FIVE-YEAR-OLD CHILDREN

Children cannot make drawings of things or people until they are five years old. Before age five they "scribble"—producing a drawing that represents something to the child

6.4 Drawing by Richard Eberhard, age 5.

that might not be recognizable to adults. And then, at around five years of age, a child can draw simple shapes like rectangles and triangles in two dimensions, but they are not able to draw circles or three-dimensional objects.

When a child around age five is asked to draw a house, it is almost certain to look like the one in figure 6.4. Rhoda Kellogg, who collected children's drawings from all around the world, shows 2,951 drawings of simple façades that look remarkably similar, regardless of differences in culture or construction type.[4] This seems hard to believe. Most people assume that children must be making copies of what they see in books, or of what another child as done, or of what an adult has told them to do. But in fact, this similarity is not a matter of copying another drawing, but one that seems to be innate.

The drawing in figure 6.5 was made by a child living in Spain after the Spanish Civil War. At the time, it was believed that having children make drawings was a form of therapy for dealing with the horrors of war. This drawing was first published in 1938 by the Spanish Welfare Association of America.

The drawing in figure 6.6 was made in 2001 by a child living in a small village in Mozambique. The five-year-old child, whose name was Erica Lagos, had of course never seen the 1938 drawing, but even the location of the windows is similar.

6.5 This drawing was made by a child living in Spain after the Spanish Civil War, and was first published in 1938 by the Spanish Welfare Association of America.

6.6 Drawing by Erica Lagos, Mozambique, age 5.

We must ask ourselves: why would this happen? The only reasonable explanation would seem to be that images are genetically wired in the infant brain—even before we are born. A story that might help explain this involves a thirty-four-year-old woman who suffered irreversible brain damage from exposure to carbon monoxide. She was unable to recognize faces of relatives or identify the visual forms of common object. Her damage was exclusively visual, because she could still recognize family and friends by their voices and objects by feeling them. Ten years after her accident, she participated in an experiment in which she was shown a drawing of an apple and a book, like those shown in figure 6.7, but she was unable to identify them. When asked to make a copy of the drawings, she could only scribble. Her drawing of an apple is on the left and a book on the right. As you can see, she could control the action of a pencil but could not copy visually. When asked to draw an apple and a book from memory, she produced the drawings on the left. Her deficit seems to have been "perceptual" rather than "sensory" in nature.

This experiment showed that she had access to what is called "low-level sensory information" which includes basic, familiar images such as an apple or a book. It seems

6.7 Drawings by a 34-year-old woman who suffered irreversible brain damage from exposure to carbon monoxide.

possible that children who are five have a low-level sensory information image of a house which they are able to call upon when asked to draw one, even though they cannot copy a drawing of this complexity.

OPTOGENETICS—THE LATEST METHOD OF STUDYING THE BRAIN

Although scientists have pursued the idea of using light to control the activity of cells for several decades, the traditional methods had drawbacks that limited their effectiveness. In 2010 a new method—called optogenetics—was developed. This method has the capacity to control cellular behaviors using light and genetically encoded light-sensitive proteins. Genetic targeting provides exquisite specificity in live animals in a manner that cannot be achieved with other methods. Optogenetics has already changed the way many neuroscience studies are performed. It makes it possible, for example, to study how activity in specific neurons can control specific behaviors. Thus if a neuron in an animal's brain is exposed to light which activates light-sensitive proteins that have been genetically coded for fear or joy, it is possible to observe whether either of those behaviors will result—once again linking neuroscience and architecture. Optogenetics is an example of how a technology, born as an idea in the minds of many scientists, becomes the fruit of multidisciplinary developments through collaboration.[5] Architects interested in research need to join such interdisciplinary teams both for potential intellectual enrichment and for applications in intelligent buildings.

THE EARLIEST ARCHITECTURAL SETTINGS: NEONATAL INTENSIVE CARE UNIT DESIGN

A clear example of how knowledge from neuroscience can, and should, change the design decisions made by architects is in the design of special places in most hospitals called NICUs—the intensive care units that specialize in the treatment of ill or premature newborn infants. Prior to the Industrial Revolution, premature and ill infants were born and cared for at home and either lived or died without medical intervention. The NICU environment provides challenges as well as benefits. The protected microenvironment includes continual bright lights, a high noise level, reduced physical contact, and painful procedures that unduly stress the infants.

From the early years, it was reported that children who were cared for in NICUs grew up with a higher proportion of disabilities, including cerebral palsy and learning difficulties, than normal children. Now that treatments are available for many of the problems faced by tiny or immature babies in the first weeks of life, long-term follow-up, and minimizing long-term disability, are major research areas. Dr. Stanley Graven at the Department of Community and Family Health, University of Florida, has described in some detail the impact of light and noise on the fetus and on children born prematurely. From his knowledge of neuroscience as well as pediatrics he is able to critically appraise the architectural design decisions of traditional NICUs. His basic concept is to design an NICU to respond to the needs of sensory systems of the fetus and premature child rather than the doctors and nurses. Design decisions need to support and facilitate development, and minimize interference.[6]

The most important neurodevelopment in the early stage of the life of a fetus is the basic structure of the brain; the development of the nerve tracks, the development of the sensory organs and the basic connections and pathways that connect them to one another. After this first stage of development has been completed, the basic structure of the eyes and ears, with their pathways into the central nuclei and then to the cortex, is genetically driven. What vision and hearing will be capable of responding to later in life is influenced by the stimuli that the baby receives during certain periods in their development—called critical periods. In the third stage of development, the processes are almost entirely driven by responses to stimulation from the architectural setting—that is, either from the mother's womb or from its surrogate created in the NICU. Because new pathways, memory circuits, and the whole range of new connections for the neurons in the cortex are made in this stage, it is crucial that the architectural setting responds to the child—more so than to the doctors and nurses whose imperatives set the traditional criteria for NICU design decisions.

The critical period in development, mentioned above, is based on the biological fact that there is a specific window of time in which external stimulation results in the correct placement and relationship of neurons in the cortex. This is especially important during periods of visual and auditory development. The critical auditory period begins about eight weeks before birth and continues through the first and second years of life. These early months are the most important period when auditory stimuli result in changes in the auditory nucleus—it is important because during this time frequency discrimination and pattern recognition develop. Interference with hearing systems during this critical period can result in a decreased ability to discriminate frequencies and to recognize sound patterns. This can mean that the child will never be able to develop perfect pitch, which is important to musicians. The critical period in visual development is initiated with the first visual exposure after birth. Combinations of light and images initiate a biochemical process which stimulates the neurons of the optic cortex. During the first five or six months after birth the infant develops the neuron relationships that allow him or her to perceive lines and shapes in subsequent life. And, as indicated in the discussion above of house drawings by five-year-old children, this may be when the classic image of a house is formed. This is called a critical period because the movement of neurons cannot occur after this period has passed.

There are a number of instances of interference with these critical stages, but the most common example is the early introduction of visual stimuli before auditory patterns are learned and in place. This can create serious concerns for chemosensory, auditory, and visual development in the premature infant setting in the NICU because many of the stimuli are either at inappropriate levels or out of sequence for the normal development processes. For example, light is a source of energy and is capable of producing injury under a variety of circumstances. The risk of damage from light is related to the wavelength, the intensity, the duration, characteristics of the eye, and the maturation of the eye and the eyelids, etc. While the eyes can tolerate brief periods of intense light, the infant's eyes also need periods of very reduced light for retinal regeneration. There are examples of NICUs that were designed with windows to provide a view for the nurses and doctors, as well as visiting family. While they were well-intentioned, such design decisions were made without knowing that the premature infant's eyelid cannot protect its visual system from natural light, which is *10,000* times brighter than electric lighting.

Much the same is true of noises. Infants exposed to constant high levels of background noise from cooling equipment, air-handling units, and communications systems (e.g. "Dr. Kildare is wanted on the third floor" is broadcast over loudspeakers in the nursery) have been shown to have major long-term negative effects. Under noisy conditions the bandwidth for the reception of sound in the ear actually increases or widens so that the infant, as a child and as an adult, will be less able to discriminate between frequencies.

OPEN AWARENESS

A form of attentiveness characterized by a receptivity to whatever floats into the mind is known as open awareness. Experiments suggest this kind of attention is the source of our most creative thoughts.[7] Open awareness frees the brain to make chance associations that lead to fresh insights. Architects and neuroscientists alike seem unusually adept at such productive daydreaming.

We tend to mistakenly think of attention as a switch that is either on or off. But instead, attention comes in many varieties. If we are too attentive, we tend to suffer from tunnel vision—the mind narrows. On the other hand, if we are not attentive enough, we lose control of our thoughts—we become scatterbrains. In between the two is a state where we are using our "double helix" of neuroscience and architecture; we are in a world of happiness (a pleasant mood of reflection) and pensiveness (hoping something good will soon reach our attention).

All forms of attention arise from the interplay between two different parts of the brain. The older, lower brain, working largely outside of consciousness, constantly monitors the signals coming in from the senses, making us aware of shifts in our surroundings—we are no longer outside in a park but in our laboratory or drafting room, for example—or causing us to remember things we are worrying about, such as getting paid. This form of attention—called "bottom-up" attention by neuroscientists—is impulsive, uncontrolled, and often fear-driven. When these sorts of signal reach our conscious brain, we tend to "turn our attention" to them. The neocortex, on the other hand, is the brain's more recently evolved outer layer that works to control these primitive impulses. The source of such "top-down" attention is what enables us to screen out distractions and focus our mind on a single task. Goleman warns that smartphones and other network gadgets force us to practice "the impoverishment of attention," and consequently cause us to become prisoners of our bottom-up attention circuits.

BACK TO THE BEGINNING

The Academy of Neuroscience for Architecture was originally formed in 2002 as a "legacy" project of the San Diego chapter of the AIA. The intention was to create something that would continue to exist beyond the 2003 AIA convention that was to be held in San Diego. This organizing effort was undertaken by a small group of architects and neuroscientists who eventually formed ANFA and became its founding board members. An important part of this effort was to enlist "Rusty" Gage (Dr. Fred Gage from the Salk Institute's Laboratory of Genetics) to deliver the keynote address to the AIA Convention. His summary of his thoughts provides a fitting summary to this essay:

> The brain controls our behavior.
> Genes control the blueprint for the design and structure of the brain.
> The environment can modulate the function of genes, and
> ultimately the structure of our brains.
> Changes in the environment change the brain, and therefore
> they change our behavior.
> Consequently architectural design changes our brain and our behavior.[8]

Ten years ago, when I first became aware of the potential for finding applications of neuroscience to architectural problem-solving, there were only a few professionals interested in such studies. Today there are both neuroscientists and architects around the world who are exploring this potential. A recent workshop organized by the Academy of Neuroscience for Architecture, for example, was attended by ninety people from twenty different countries. While there is little actual new knowledge available so far, momentum is gathering toward a promising future.

NOTES

1. Formalin is a saturated solution of formaldehyde dissolved in water with another agent like methanol.

2. Patricia Churchland, *Brain-Wise: Studies in Neurophilosophy* (Cambridge, MA: MIT Press, 2002).

3. Text is available under the Creative Commons Attribution-ShareAlike License.

4. Rhoda Kellogg, *Analyzing Children's Art* (Palo Alto: Mayfield Publishing, 1970).

5. For further information, see Optogenetics Resource Center, Stanford University, Palo Alto.

6. Stanley Graven, "Early Visual Development: Implication for the Neonatal Intensive Care Unit and Care," *Clinical Perinatal* 38 (2011): 621–684.

7. Daniel Goleman, *FOCUS, The Hidden Driver of Excellence* (San Francisco: Harper, 2013).

8. Rusty Gage, keynote lecture at the 2003 AIA National Convention, San Diego, 10 May 2003.

7

NESTED BODIES

Sarah Robinson

Change your brain, your body, your environment in nontrivial ways, and you will change how you experience your world, what things are meaningful to you, and even who you are.[1]

Mark L. Johnson

We are bodies who start inside other bodies. Most of us think we know what our body is: it is the fleshy whole we inhabit; but the dictionary defines the word "body" much more broadly. A body is the entire material or physical structure of an individual organism; it is also an entity composed of numerous members—of people, things, concepts, or processes—a student body, a body of work, a body of evidence, the body politic. Body is used to describe the main or central part of something—the body of a temple, for instance. It can also describe a mass as distinct from other masses—a body of water or a celestial body. Body can also be used to describe a qualitative measure of physical consistency; wine and sauces have a certain body. Shakespeare used body as a verb, "To body forth the forms of things unknown." What does "body" mean? In some, but not all, of these meanings, body is used to refer to a material entity. What all the meanings do share is the sense that the body is a boundary that delimits qualities, persons, ideas, substances, objects or processes. Yet even this widened definition is tenuous and provisional at best. The line where our body begins and where it ends is the subject of controversy in numerous disciplines. From the perspective of physics, we know that energy fields are unbounded. The biomagnetic field of the human heart extends indefinitely into space. The strength of the field may diminish as it moves farther from its source, but there is no definite point at which you can say the field ends. Breakthroughs in quantum physics have made it possible to develop instruments so sensitive that they can detect the biomagnetic field of the human heart from 15 feet away.[2] So, whenever we share the company of others, we find ourselves in the midst of overlapping, interpenetrating bioelectric and magnetic fields that originate within each of our bodies.

Defining our bodies as distinct entities composed of matter is similarly contingent and marginal. In physics, there are two worlds: the classical world and the quantum world—particles that comprise matter behave differently in each of these worlds. An electron is able to tunnel through a material in one world, but not in the other. In the classical world, electrons behave as particles and in the quantum world they behave as waves. Neither world, classical nor quantum, can be understood by drawing hard lines: energy fields extend indefinitely, matter can be reduced infinitely.

The body also includes the mind. Our mental activities originate within and are beholden to the body. Each of our conscious states and every cognitive function—emotion, thought, perception, desire, memory, imagining—is generated, in part, by galaxies of electrochemical interactions that take place in the body. Neuroscientists developed the concept of "neural correlates" to link first-person subjective states to the biological activity that generates them. The increasing sensitivity of fMRI scans has permitted researchers to begin to trace the physiological labyrinth that elicits the human experience of love, hate, compassion, and attention.[3]

Neuroscientists continuously expand their vocabulary in order to discuss the complex weave of interrelated activities that occur within and because of our bodies. Over a century ago, they introduced the "body schema," a notion that refers to the body's relations to immediately surrounding space. The body schema includes the brain and sensory processes that register the posture of one's body in space. The body schema is plastic, amenable to constant revision, extends beyond the envelope of the skin, and has important implications for tool use. Recent studies have shown that tools are incorporated into the body schema within seconds, regardless of whether the subjects of the experiment had prior training or exposure to them. Our body readily integrates tools into its organized model of itself.[4]

We can no longer consider the organism and the environment to be independent entities. While the body schema refers to the area immediately surrounding the body, neuroscientists use two terms to describe the space beyond the body schema: peripersonal space

7.1 Friedrich Hundertwasser, *Arnal—Mishap I—The Suffering of Mishaps in Love*, 1965.

describes the space immediately surrounding our bodies; extrapersonal space refers to the space just beyond the peripersonal. The line drawn between these layers is artificial, of course; its use facilitates more precise study. The brain perceives objects situated in different positions of space through multisensory information that is integrated in multiple areas of the brain and body. The body schema, peripersonal space, and extrapersonal space, rather than being distinct entities, are emergent attributes of interacting cortical and subcortical areas.[5] In other words, our body's apprehension of surrounding space and its contents comes into being through a dynamic, multisensory process irreducible to a gross measure of inside and outside.

The consensus among experts in biology, psychology, cognitive neuroscience, and philosophy is that none of our experience, thought, and communication would exist without our brains functioning as organic members of our bodies, which in turn are actively engaged with the specific physical, social, and cultural environments in which we dwell. "Organisms and environments are co-evolving aspects of the experiential processes that make up situations," writes the philosopher Mark Johnson.[6] The philosopher Alva Noë, in venturing to predict the trajectory of future research in neuroscience, writes: "Just as we do not draw an impermeable boundary around the brain, we will not draw such a boundary around the individual organism itself. The environment of the organism will include not only the physical environment, but ... the cultural habitat of the organism."[7]

Researchers in the diverse fields of cognitive science, anthropology, evolutionary biology, psychology, and philosophy, among others, also converge on their appraisal of human social intelligence. They broadly agree not only that human cognition evolved for sophisticated communication between individuals, but that cognition is also socially distributed. "Individuals have evolved to live in groups, not in isolation," writes the neurophysiologist Walter Freeman.[8] Language, for instance, can be considered as the collective, decentralized product of cognition—an accretion of human knowledge invented by no one that belongs to everyone. Cultural artifacts and institutions, including architecture, sediment aspects of shared knowledge and meaning as objective features of the world.[9] Without these social and cultural enterprises, our cache of collective knowledge and values would not be preserved over time, and each new generation would suffer amnesia—and be forced continually to start anew.

A truly up-to-date understanding of the body must include the mind and all of its processes: desire, emotion, cognition, remembering. Our body also includes the tools with which it extends itself. The walking stick in the hand or the violin beneath the chin are each—according to the brain—extensions of itself. A comprehensive understanding of the body must also encompass the environments in which we interact and on which we depend. Though it is empirically responsible and technically accurate, understanding the body with such complexity and depth is a radical departure from the reigning paradigms of Western thought.

Indeed, coming to terms with the implications of our embodiment is "One of the most profound philosophical tasks you will ever face. Acknowledging that every aspect of human being is grounded in specific forms of bodily engagement with an environment requires a far-reaching rethinking of who and what we are,"[10] writes Johnson. The breakthroughs in the cognitive and neurosciences have forced researchers in diverse disciplines to reorient their inquiries toward understanding the bodily basis of human meaning. This shift away from the disembodied mind of an isolated individual, toward the incarnation of meaning through the interaction of embodied beings actively engaged in their environments and with each other, raises the stakes for the architect. If the philosopher's role is to fathom the conceptual world, ours is to build its material manifestation.

A LONG-AWAITED PARADIGM SHIFT

The history of science is a series of updates, revisions, and expansions of earlier models of the universe. When the prevailing model fails to provide internal consistency with the introduction of new experimental data, one is faced with two alternatives: either to deny the validity of the observations, or to expand the model sufficiently to allow for their natural inclusion. The discovery of quantum phenomena forced physicists to amend their model of the universe. Over the last three decades, the body of evidence amassed in the sciences devoted to studying the mind has brought us to a similar juncture in architectural practice.

As architects, we ostensibly shape matter—which sometimes behaves like a particle, at other times like a wave. Yet, like classical physicists, we tend to address the dimension of our work that behaves as if it were a particle. The subtler dimensions, the layers that engage emotions, provoke imagination, empathy and social contact, tend to be invisible, irreducible, and therefore undervalued, overlooked, and even denied. Considering the material and physique of our work in the enriched and nuanced way that we now must understand the human body would go a long way toward restoring meaning and relevance to architecture. Our impoverished sense of matter has partially contributed to the sense that architecture is one among many consumer goods. The meaning and presence of a building does not stop at the surface of its skin any more than your body does.

Buildings are extensions of our bodies in profound and pervasive ways. To begin to understand the extent to which architecture interacts with our bodies, let us consider the body as a metaphor for architecture. We now know that we must select our metaphors with extreme care.[11] Paradigm shifts in the sciences have been precipitated by shifts in the selection of metaphors. Cognitive scientists tell us that metaphors are an innate feature of our conceptual system; we simply cannot think without their support. Metaphors are not psychologically or ideologically neutral, they open epistemic channels of associations and meanings while closing off others. For decades the cell was characterized as a bag of empty space with billiard balls bumping around its interior. We now know that the cell is so packed with filaments, tubes, and fibers—collectively called the intracellular matrix—that there is little room for a solution of randomly diffusing objects.

In architecture we must also introduce metaphors more appropriate to the real enterprise in which we are engaged. In shaping matter, we shape experience—in shaping experience, we give form to life. The language and metaphors we use to communicate and imagine our work are in urgent need of an overhaul. Conceiving the building as an inert object has the validity of Newtonian physics: it is useful in circumscribed cases, but its relevance is consigned to history. The metaphor of the body opens a more complex and subtilized understanding of architectural potentiality.

NESTED BODIES AND THE SENSES

Like our body, a building is a series of interrelated systems, each possessing its own identity and offering a particular array of affordances. The mind is nested in the body and the body is nested within the contexts of room, building, city, earth, universe. We could say that our body has, nested within it, at least four bodies: our physical body, and the more ephemeral, but equally real, emotional body, mental body, and social body. The action of these bodies is expressed through our perceptual systems. J. J. Gibson identified our five modes of perception, which he called externally oriented attention, as the basic orienting system, the auditory system, the haptic system, the tasting-smelling system, and the visual system. If we lay these five modes of attention over each of our four bodies, we can begin to develop a framework for exploring how the body interacts within architectural settings.

This exercise highlights aspects of architecture that can be identified as specific challenges for studies in the cognitive and neurosciences. The insights yielded from such investigations would help architects design in a manner that more fully engages and supports our bodies, minds, and social and cultural evolution. The findings of such experimental work would not only benefit architects, but would advance studies in neuroscience as well. Human behavior is always situated, and cannot be fully understood apart from its context. And since we spend over 90 percent of our lives in buildings, studying human response within the context of buildings would elucidate our most pervasive reality. Research in the cognitive and neurosciences could invigorate and expand our understanding of what architecture is, and of what it can do and be.

SPATIAL COGNITION AND SENSE OF PLACE

Gibson's first category of external attention, the "Basic Orienting System," is the basic frame of reference for all the other perceptual systems. This system comprises the vestibular organs in the inner ear that function in harmony with our eyes and our

7.2 Grimshaw Architects, Via Verde housing project, Bronx, New York: integrated roof gardens.

proprioceptive sense of weight to organize our world into vertical and horizontal planes. Spatial cognition, our awareness of the surrounding environment, plays a fundamental role in all of our behavior. John O'Keefe and his colleagues at the University College in London were among the first to identify "place-cells" and their role in creating spatial maps.[12] In their studies of rats, they discovered that an internal map of space develops within minutes of the rat's entrance into a new environment. Like our capacity to learn language, the capability to build a spatial map is innate, but the particular features of the map depend on each person's experience.

Spatial cognition, an inherent expression of our basic orienting system, is wrought with emotional, mental, and social implications. We now know that the architecture of each person's brain is unique, and its uniqueness stems in part from the places we experience.[13] What makes a place work its way into our psyche? Why do we love and vividly remember certain places and not others? Can we remember *anything* outside the context of place? How do we create places that people will care for and cherish?

The Via Verde mixed-use housing project in the South Bronx area of New York City exemplifies the holistic approach warranted by a more sophisticated appreciation of the interaction between mind, body, and place. When asked what kind of building they wanted, the neighbors and future residents told the design team that they simply wanted a healthy place to live. To meet this deceptively straightforward request, the interdisciplinary team enlisted a group of healthcare professionals to locate their clinic in one of the retail spaces. The premium location, with the best views, was given not to an exclusive residence, but to the communally accessed fitness center. Questions of health have far-reaching lifestyle implications that are central to good design and sustainability. Michael Kimmelman, reviewing the project for the *New York Times*, observed: "The greenest and most economical architecture is ultimately the architecture that is preserved because it is cherished. Bad designs, demolished after 20 years, as so many ill-conceived housing projects have been, are the costliest propositions in the end."[14]

AUDITORY SYSTEM AND ACOUSTIC ARCHITECTURE

Gibson's second category of externally oriented attention is the auditory system which orients us to sound and airborne vibrations. Investigating the architecture of sound and echo would shift design practice away from its overemphasis on the visual. Blind people are connoisseurs of sound and echo. Designing with an awareness of their refined spatial awareness would open a new dimension of architectural experience for sighted people, as well. As Iain McGilchrist has pointed out, John Hull, in his book *Touching the Rock*, offers profound insights into the multisensory experience of space from his own experience of blindness. In one instance, he notices that for sighted people, the affability of the weather is determined by predominately visual criteria. A nice day is synonymous with a clear blue sky. The blind person evaluates the climate differently. Hull writes: "For me, the wind has taken the place of the sun, and a nice day is a day when there is a mild breeze. This brings into life all the sounds in the environment. The leaves are rustling, bits of paper are blowing along the pavement, the walls and corners of the large buildings stand out under the impact of the wind, which I feel in my hair, on my face and in my clothes. A day in which it is merely warm would, I suppose, be quite a nice day but thunder makes it more exciting, because it suddenly gives a sense of space and distance.

Thunder puts a roof over my head, a very high, vaulted ceiling with tumbling sound. I realize that I am in a big place, whereas before there was nothing there at all. The sighted person always has a roof overhead, in the form of a blue sky, or clouds, or the stars at night. The same is true for the blind person of the sound of the wind in the trees. It *creates* trees; one is surrounded by trees whereas before there was nothing."[15]

For the blind person, the sound of the wind creates a tree and thunder builds a roof. Such a nuanced description of the spatial properties animated by the wind helps one appreciate the extent to which the acoustics of the air have emotional, mental, and social resonances. Airborne sound impacts concentration, one's sense of privacy as well as one's fantasies of freedom and flight. Researchers who study the buildings of ancient cultures suggest that their design may have been generated from predominantly acoustical rather than visual considerations. The caves at Lascaux may have been adapted to mimic the sound of beating hooves, Mayan pyramids may have been designed to sound like rain—these ancient places relied on human social interaction and participation to activate the multisensory space.[16] Just as the tree comes alive in a thunderstorm, so too can the actions of people become design features that enliven architectural space.

THE HAPTIC SYSTEM AND TEXTURE

Gibson's third category of externally oriented attention is the haptic system, the system which pertains to our sense of touch. Touch does not stop at the skin; it involves deformations of the tissues, configurations of joints, and the stretching of muscle fibers through contact with the earth. In the haptic system, the hands and other body members are active organs of perception. Touch receptors in the skin combine in the brain with vestibular, visual, and other touch information to help us retain our equilibrium. Studies show that Europeans who walk on cobblestone lose their balance more slowly than Americans who walk on flat sidewalks.[17]

Stone-stepping in Taiwan is an ancient practice known to stimulate the touch receptors in the feet, an effect that percolates through the entire body. The floors of ancient buildings document human movement. The marble floor of the Hagia Sophia, for instance, is a sentient film etched with nearly fifteen centuries of human motion. Footsteps in such places eventually melt hard stone.

Our haptic sense is linked with our emotional and mental bodies; crossing the threshold of a Wrightian entrance imparts a sense of compression—a necessary prelude to its counterpoint, the exhilaration of release. Such a passage engages the body in a movement equivalent to an orchestral movement in a symphony. The resistance encountered when opening a heavy door imparts the sense of gravity, formality, and pause before entering into another realm. A sense of repose and relaxation is fostered in refuge-type settings, whereas prospect-rich circumstances beckon the imagination.

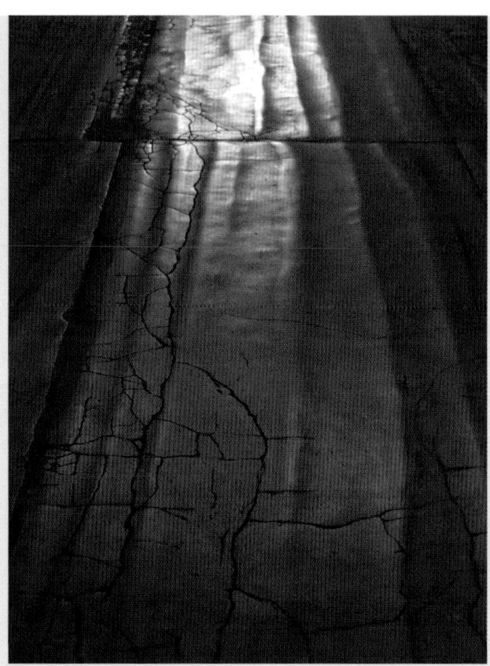

7.3 Worn pavement, Hagia Sophia, Istanbul.

Texture invites life. Consider the undulating curves of Studio Gang's 82-story Aqua tower, for instance. The curves break up the notorious Chicago wind so effectively that the architects could dispense with the "tuned mass damper," a contraption that typically weighs hundreds of tons and counteracts the force of the wind. Modulating an otherwise flat, glass façade with texture renders the building more sympathetic to the migratory path of birds.[18] Because the wind is treated as a hostile force for tall buildings, most of them lack balconies. But on the Aqua tower, the large-scale texture of the curves creates a vertical landscape that shelters balconies in quiet eddies. The syncopated curves create opportunities to interact with one's neighbor. Instead of meeting in the hallway, people's eyes first meet on their balconies. According to Gang, many love affairs between residents were started in just this way.[19]

Our ability to understand others—empathy—is the foundation of human relationship, social intelligence, and cohesion. Vittorio Gallese and his team in Parma, Italy, have studied the relationship between touch and social cognition. According to Gallese, through embodied simulation we make sense of others' experiences by activating the same neural

7.4 Frank Lloyd Wright, Fallingwater, Mill Run, Pennsylvania: balanced between prospect and refuge.

structures that are involved in our own bodily experiences. We do not understand others analogically through sight and inference. Rather, vision, touch, and action are inextricably linked, so that when we perceive the tactile experience of others, our own motor and somatosensory systems are systematically activated.[20] We do not understand others by merely seeing them and inferring their mental state, but through feeling what the other person is experiencing in our own bodies.[21] In this transaction we are literally "touched" by the other person.

This nascent receptivity is not limited to animate beings, but extends to inanimate things as well.[22] Somatosensory cortex function has also been linked to empathic ability.[23] If, in

7.5 Studio Gang Architects, Aqua tower, Chicago: large-scale texture.

fact, our sense of touch is implicated in our capacity for empathy and, by extension, in social perception, is it not imperative to design for human touch?

The senses can be refined with practice. The Bauhaus conducted experiments that developed tactile awareness and gave the sense of a musical scale of textural values.[24] The Japanese are known for their textural intelligence. We are makers of the material world—what are we doing to develop our own sense of touch? How can we make buildings that are irresistible to touch?

7.6 Studio Gang Architects, Aqua tower, Chicago: balconies as sheltered eddies.

TASTING, SMELLING, AND THE IMAGINATION

Gibson's fourth category of externally oriented attention is the tasting-smelling system. The body's oldest perceptual systems, the olfactory and gustatory systems alert us to perils in the environment. This perceptual system is rooted in the limbic system, an area that, according to the neurophysiologist Walter Freeman, plays a central role as an organ of intentionality as it provides for the sense of time, space, and expectancy of purposes and goals.[25] We know that being in gardens and natural settings helps to heal patients in hospitals. How can we design for the biophilia of scent?

The footprints of the buildings in the Via Verde project were made intentionally narrow to allow the apartments to wrap around the central courtyard. Narrower spaces permit cross-ventilation and the deeper penetration of natural light. Fans were installed to discourage the use of air conditioning. The terraced roof is planted with garden plots and fruit trees. Light, air, and scent become free-floating design features.

Smell is intimately connected with long-term memory.[26] Memory and imagination share the same neural pathways in an interaction so entwined that the Nobel laureate neurophysiologist Gerald Edelman said that every act of memory is also an act of the imagination.[27] Smell and taste obviously contribute to social cohesion. After all, gathering around a fire was one of the preadaptations that paved the evolutionary path to becoming human.[28] Eating has complex ritualistic dimensions that are intimately related to the atmosphere in which the meal takes place. Eating together with others around a common table contributes to lower incidences of obesity and other eating disorders.[29] Cultivating one's garden in a communal urban plot is a practice that reaches far beyond one's personal well- being—to nurture social contact and pride of place.

VISUAL PERCEPTION AND CHRONOBIOLOGY

Gibson's fifth perceptual system is the visual system in which the eye, the brain, and the body function together inseparably. The visual system registers the constantly varying intensity of multidirectional light. The cells in the retina do not signal absolute levels of light; they signal the contrast between light and dark. The brain derives meaning from quickly surveying a visual setting, not by recording the scene passively like a camera would—perception is inherently creative. "Every act of perception is an act of creation,"[30] says Edelman. Understanding the dynamic nature of visual perception would help us design with the poetry of light and shadow. The most effective stimulus for retinal cells, for example, is not evenly distributed light but linear contours with elongated edges between the light and dark areas.[31] Shadows deepen mystery and ambiguity. Semir Zeki has convincingly shown that ambiguity in art activates the imagination.[32] An unfinished image calls upon our creative resources to complete its story, to collude in its plot—to round out its body.

For decades, Juhani Pallasmaa has argued against the oculocentric bias of contemporary architecture, a tendency historically rooted in the perspectival understanding of space that "gave rise to an architecture of vision, whereas the quest to liberate the eye from its perspectival fixation enables the conception of multi-perspectival, simultaneous, and atmospheric space. Perspectival space leaves us as outside observers, whereas multi-perspectival and atmospheric space and peripheral vision enclose and enfold us in their embrace."[33]

This observation resonates with John Hull's spatial experiences. Our sense of vision is so powerful that we tend to literally overlook the reality of other worlds. We mistake the world in front of us—the one we can verify with vision—for the sole reality. Peripheral vision and echolocation cause space to open out around us, shifting our relationship with the world from one of domination and control to one of inclusion and nurture.

Peripheral vision cannot perceive details with precision because it appraises a situation holistically. Perhaps this is why peripheral vision has been linked with the ability to gauge the emotional state of others. The neurophysiologist Margaret Livingstone has pointed out that Leonardo da Vinci exploited the complex nature of the human visual system in his masterpiece, the *Mona Lisa.* When you stare straight at the painting, your central vision focuses on the details of Mona Lisa's face, and you miss her enigmatic smile. Just as you must use your peripheral vision to view a distant star, if you glance at the side of Mona Lisa's face, her smile emerges quite clearly.[34]

Our eyes also contain nonvisual rods and cones that synchronize our endocrine system with larger cycles of day and night. Research in chronobiology, the science that studies circadian rhythm in living creatures, has shown that lack of daylight in buildings and the absence of darkness at night cause endocrine disruption. Endocrine disruption is known to cause hormone-related disorders such as breast and prostate cancer, infertility, precocious puberty, and insomnia.[35] We know that low light levels cause depression, and that views toward greenery evoke our innate biophilia. Colors influence emotional states. The evolutionary nature of our visual system explains why we find settings reminiscent of the savanna of our distant ancestors highly desirable, and why certain proportions are more satisfying than others.[36]

THE ATMOSPHERIC SENSE AND SPACE AS PLENUM

To fully appreciate the complex, nuanced nature of our interaction with our environment, we must, as Juhani Pallasmaa recommends, introduce a sixth sense: "Our capacity to grasp qualitative atmospheric entities of complex environmental situations, without a detailed recording and evaluation of their parts and ingredients, could well be named our sixth sense, and it is likely to be our most important sense in terms of our existence and survival."[37] The realm of the atmospheric sense is at once fine and fundamental, and perhaps best described by someone with special abilities. John Hull's narration of the experience of rain betrays his highly developed atmospheric sense: "Rain has a way of bringing out the contours of everything; it throws a colored blanket over previously invisible things; instead of an intermittent and fragmented world, the steadily falling rain brings a continuity of acoustic experience. ... I hear the rain pattering on the roof above me, dripping down the walls to my left and right, splashing from the drainpipe at ground level on my left, while further over to the left, there is a lighter patch as the rain falls almost inaudibly upon a large leafy shrub. On the right it is drumming with a deeper, steadier sound, upon the lawn. I can even make out the contours of the lawn, which rises to the right in a little hill. ... The whole scene is much more differentiated than I have been able to describe because everywhere are little breaks in the patterns, obstructions, projections, where some slight interruption or difference of texture or of echo gives an additional detail or dimension to the scene."[38]

Imagine sensing the presence of a small hill without the convenience of vision! Hull's description suggests that space itself is full of texture and pressures and currents. Here the rain "bodies forth" the inchoate topography of space. Hull does not consider himself to be blind, but rather identifies himself as a whole- body seer. This panoramic, total-body perception is the *sine qua non* of the atmospheric sense. Being immersed in the atmospheric moods provoked by the weather has a profound impact on our mental and physical well-being, for instance. We can all relate to Marcel Proust when he says: "A change in the weather is sufficient to re-create the world and ourselves." The atmospheric sense subsumes the other five senses—our entire being—in an ambient awareness of space as a plenum.

The interior of the cell was once thought to be empty, but we now know it is filled with the intracellular matrix of filaments and fibers. The notion of absolute space, like the hard line that separates our body from the rest of the world, was created for theoretical convenience. Space is not empty; it is latent with possibility. The atmospheric sense comes alive in our conscious immersion in the omnipresent rhythms of life.

THE LIMITS OF QUANTIFICATION

Truly reckoning with our embodiment means we must recognize that we are situated in a world whose subtle relations and intricate causality pattern our being at every level. Our advanced technology permits us to delve into the smallest and largest scales imaginable—what we find is at each pole is infinity. The human brain alone has over one hundred billion neurons, and each one has an average of 7,000 synaptic connections, which means we each have roughly 500 trillion synapses. And let us not forget that the brain is but one member of the nervous system that extends from the ends of our fingers to the tips of our toes.

The tools we use to augment perception permit us to investigate extreme, rarefied conditions—from quantum states to intergalactic space. But the homeostatic bandwidth of daily life does not lend itself to the same scrutiny. Research in neuroscience remained highly speculative until the advent of magnetic resonance imaging—the technology that has precipitated its current apotheosis. The annual meeting of the Society for Neuroscience attracts around 30,000 participants and generates roughly 15,000 talks and posters.[39] How does one begin to critically assess such a staggering deluge of information?

Also, it is important to bear in mind that when neuroscientists study cognitive functioning in a brain scanner, it is not the process of cognition that they directly witness. Because neural events consume oxygen, they require blood. The fMRI builds an image from the radio and light waves emitted in this metabolic exchange. Brain scans represent mental activity three steps removed from the actual cognitive process; first they quantify the physical magnitude correlated to the blood flow. The blood flow is then correlated with neural activity. And finally, if all of these inferences are correct, a brain scan might reveal important information about the neural activity related to a cognitive process.[40]

Brain scanners have relatively low spatial and temporal resolution. They allow scientists to localize neural activity to cubic regions of between 2 and 5 millimeters: an area that contains hundreds of thousands of cells. In an area of this size, specialization or differentiation between the cells will not appear on the resulting image. Cellular activity occurs at scales of thousandths of a second, but the scanner requires time scales of fractions of a minute to detect and process signals for making images. For this reason, neuroscientists have developed techniques for averaging these test results. In this normalizing process, significant data are inevitably lost. If our brains are as unique as our fingerprints, how do we determine what is average? The colorful images of brain activity found in science journals are not depictions of an actual person's brain, but rather normalized findings projected onto a hypothetical stock brain.[41]

These methodological shortcomings underline the fact that neuroscience, like the accrual of all human knowledge, is vulnerable to error, misconception, and conceit. Knowledge advances in tandem among multiple disciplines that share and temper each other's findings. Neuroscience studies the nervous system, which is situated in the body, which is in turn situated in an environment; no single discipline can work on the sheer complexity and vastness of such a project alone.

COSMOLOGY OF THE NEST

"Our architecture reflects us as truly as a mirror, even if we consider it apart from us."[42] wrote Louis Sullivan. Every architecture embodies the mental orientation of its makers. Contemporary architecture is an artifact of the philosophical and scientific inheritance of the Enlightenment.[43] The Age of Reason supplanted the worldview that governed most of Europe until the seventeenth century. In this earlier Ptolemaic and Aristotelian cosmology, the universe was understood as a series of concentric, transparent, hollow spheres that orbited independently of one another around the earth. The surface of each crystalline orb supported one of the planets. In this model, the universe was conceived to be a series of nests that held bodies—heavenly bodies.

The Egyptians and Mesopotamians shared this intimate, protected conception of the world. They considered the sky to be a vast tent, pitched across the peaks of the highest mountains. The architecture of native peoples throughout the world similarly embodied this sense of the world as a massive interior. For the Sioux Indians, the walls of their teepee were skin and the supports were bones; in their leaning one upon another they formed a soft, giant body. The Navajo hogan is a cosmos in miniature. Similarly, the Haida tribe modeled their traditional cedar-plank houses as microcosms of the wider world. This architecture was the natural extension of their bodily sense of themselves. Self was not isolated and separate, but related, interpenetrating, interdependent—not bounded by a hard shell, but porous and inclusive.

The heliocentric model effectively turned the world inside out. This exteriorization of the universe had profound psychological and philosophical consequences. The philosopher and ecologist David Abrams suggests that the dissolution of the pre-Copernican cosmos exiled the qualities, feelings, and animating spirits once felt to belong to the surrounding terrain to the subjectivity of each person's inner world. The once protected cosmos became a boundless world of objects obeying mechanical laws. This new version of the world could no longer contain the felt relations between creatures and places. "These ambiguous, ever-shifting qualities quit the open exteriority of the physical surroundings, taking refuge within the new interiority of each person's 'inner world.' Henceforth they would be construed as merely subjective phenomena,"[44] writes Abrams.

This brings to mind John Dewey's controversial claim, mentioned earlier in this volume by Mark Johnson, that the feelings we experience should not be merely attributed to a subjective inner state, but ought to be considered as features belonging to the objective situation. So, instead of saying "I am sad," we should say "the situation is sad." What Dewey understood, and what neuroscience now confirms, is that "the proper locus of the affective is the entire cycle of organism-environment interaction, and not just the internal states of the organism."[45] We now know that the thoughts and feelings that populate our subjective reality are not abstractions belonging solely to us; rather, they are constantly forming patterns of experiential interaction emerging from our continual engagement with the environment.[46] What we understand to be our subjective reality is in fact an experiential process that is *in* and *of* the world, and not merely *about* the world.

AN ALTERNATIVE MODEL

In his seminal book *Architecture and the Crisis of Modern Science*, Alberto Pérez-Gómez demonstrated that the conceptual underpinnings of Western architecture were present long before technology was capable of making them modern.[47] Some observers have suggested that the revolution taking place in the neuro- and cognitive sciences rivals the Galilean revolution in physics and the Darwinian revolution in biology. To compare this shift to the cataclysm incurred by the transition from the geocentric to the heliocentric cosmos is quite a radical claim. If this is indeed a revolution, it is one that reveals our identity as creatures of relationship, utterly dependent on our surroundings for our physical and mental existence. Galileo peeled back the limits of the known world—in a triumph of the intellect shattering its own boundaries. This latest revolution also fundamentally redefines our boundaries. The current paradigm shift has the potential to restore our position in a cosmos whose balance depends on understanding and honoring the boundaries imposed by the exigencies of relationship—a sense of relatedness that naturally extends to the built world.

Relationships not only limit, they also allow. Iain McGilchrist has pointed out that baby birds are as attached to their nests as they are to their mothers. John Paul Eberhard has shown how the architectural setting of the NICU is the primary caregiver for a premature infant. In both cases, the environment is a primary agent in the survival and flourishing of the vulnerable creature. Acknowledging the profound agency of architectural and environmental settings is a radical departure from conventional understanding, yet it is exactly the shift necessitated by the findings of the sciences that study the mind.

In his prescient postwar manifesto *Survival through Design*, Richard Neutra wrote: "Neither physically nor biochemically nor sociologically can the individual really be segregated as a separate entity ... the organism is chemically so united with the environment that the two can only be separated in the abstract way in which we separate the water of two tributaries which have flowed together in a common river bed. Organisms are immersed to fusion in their chemical as well as their social setting, they literally live on and in one another. The isolation of the individual from his fellows is neither a biochemical nor a social fact."[48] The biological sciences force us to recognize the fundamental unity between ourselves and our environment. Our immediate environment is fabricated by us; we will only continue to discover more ways that architecture structures our physical, emotional, mental and social well-being. "Everything changes according to the interacting field it enters,"[49] observed John Dewey. Our buildings are fields of potential interaction: fields that interpenetrate, intermingle, and overlap. The next generation of buildings will be active organs in interconnected ecological, cultural, and social systems. As such, they will be technologically, environmentally, and psychically sustainable. They will possess the characteristics of organisms: affording comfort, opportunity, enrichment, and challenge to the body, in scale and performance. They will acknowledge and celebrate the plurality of time in its myriad evolutionary, geologic, seasonal, historic, and hormonal expressions. They will be created with and engage all of our senses, including those we have not yet named. Providing vital sustenance to the imagination, they will be agents in human flourishing. This project will require collaboration across disciplines that stretch from the most subtle of the humanities to science's hardest core.

"Nature has neither kernel nor shell, she is everything all at once,"[50] wrote Goethe. He personified the refinement of art and the rigor of science. More than two centuries later, the sciences of the mind are finally substantiating the claims of his delicate empiricism. "Science has not and never will have the same ontological sense as the perceived world, for the simple reason that it is a determination or an explanation of that world," wrote Maurice Merleau-Ponty.[51] Science is an intensely social, cumulative enterprise, a vast repository of knowledge constantly expanding its expository power. But as Merleau-Ponty suggests, art, insofar as it is the essence of human perception and intuition, belongs to another order. Its very nature is to leap from whatever sort of knowledge the artist might encounter, and to dream a new world into being. The work of art is judged not by the binary criteria of right or wrong, but by how it moves us—how it deepens us—how it opens new dimensions in our minds and unfolds new potential in our lives. Architecture mediates our relationship with the limitless cosmos; it situates and frames the habits, passions, and potentials of human existence. Sometimes science and sometimes art, sometimes particle and at others a wave—more than a kernel or a shell, it can and must be everything: all at once.

NOTES

1. Mark Johnson, *The Meaning of the Body* (Chicago: University of Chicago Press, 2007), 83.

2. G. Bison, R. Wynands, and A. Weis, "Dynamical Mapping of the Human Cardiomagnetic Field with a Room Temperature Laser-Optical Sensor," Physics Department, University of Switzerland, *Optics Express* 11, no. 8 (April 21, 2003).

3. Walter J. Freeman, *Societies of Brains: A Study of the Neuroscience of Love and Hate* (London: Routledge, 1995).

4. Krist Vaesen, "The Cognitive Bases of Tool Use," *Behavioral and Brain Sciences* (2012): 1–17.

5. Nicholas P. Holmes and Charles Spence, "The Body Schema and Multi-Sensory Representations of Peripersonal Space," *Cognitive Process* (June 5, 2004): 94–105.

6. Johnson, *The Meaning of the Body*, 83.

7. Alva Noë, *Out of Our Heads: Why You Are Not Your Brain and Other Lessons from the Biology of Consciousness* (New York: Hill and Wang, 2009), 185.

8. Freeman, *Societies of Brains*, 12.

9. Johnson, *The Meaning of the Body*, 152.

10. Ibid., 2.

11. George Lakoff and Mark Johnson, *Metaphors We Live By* (Chicago: University of Chicago Press, 1981), 146.

12. Eric Kandel, *In Search of Memory: The Emergence of a New Science of Mind* (New York: W. W. Norton, 2006), 307.

13. Michael Arbib, lecture at "Minding Design: Neuroscience, Design Education and the Imagination," November 9, 2012.

14. Michael Kimmelman, "In a Bronx Complex, Doing Good Mixes with Looking Good," *New York Times*, September 26, 2012.

15. John Hull, *Touching the Rock: An Experience of Blindness* (New York: Vintage, 1990), 16; emphasis added.

16. Linda Geddes, "Mayans 'Played' Pyramids to Make Music for Rain God," *New Scientist* (September 22, 2009): 26–28.

17. Sarah Robinson, *Nesting: Body, Dwelling, Mind* (Richmond, CA: William Stout, 2011), 49.

18. The Aqua tower received the 2010 PETA Proggy award, which recognizes animal-friendly achievements in commerce and culture.

19. Jeanne Gang, "A Future Built with Bits and Sticks," lecture at the Chicago Humanities Festival, June 25, 2011.

20. Vittorio Gallese and Sjoerd Ebisch, "Embodied Simulation and Touch: The Sense of Touch in Social Cognition," unpublished paper, 2013.

21. Ibid.

22. S. J. Ebisch, F. Ferri, A. Salone, M. G. Perrucci, L. D'Amico, F. M. Ferro, G. L. Romani, and V. Gallese, "Differential Involvement of Somatosensory and Interoceptive Cortices during the Observation of Affective Touch," *Journal of Cognitive Neuroscience* 23 (2011): 1808–1822.

23. J. Zaki, J. Weber, N. Bolger, and K. Ochsner, "The Neural Bases of Empathic Accuracy," *Proceedings of the National Academy of Sciences* 106 (2009): 11382–11387.

24. Robinson, *Nesting*, 123.

25. Walter Freeman, "Emotion Is Essential to All Intentional Behaviors," in Mark D. Lewis and Isabel Granic, eds., *Emotion, Development and Self-Organization: Dynamic Systems Approaches to Emotional Development* (Cambridge: Cambridge University Press, 2000), 209–235.

26. Walter Freeman, *How Brains Make Up Their Minds* (New York: Columbia University Press, 2000), 32.

27. Gerald Edelman, "From Brain Dynamics to Consciousness: How Matter Becomes Imagination," Jacob Marschak Memorial Lecture, UCLA, February 18, 2005.

28. E. O. Wilson, *The Social Conquest of the Earth* (New York: Liveright, 2012).

29. Ellen van Kleef and Brian Wansink, "Dinner Rituals That Correlate with Child and Adult BMI," *Obesity* 22 (2014): E91–95.

30. Edelman, "From Brain Dynamics to Consciousness."

31. Kandel, *In Search of Memory*, 301.

32. Semir Zeki, *Inner Vision: An Exploration of Art and the Brain* (London: Oxford University Press, 1999), 22–36.

33. Juhani Pallasmaa, "Space, Place and Atmosphere: Peripheral Perception in Architectural Experience," Inaugural Kenneth Frampton Endowed Lecture, Columbia University, New York, October 19, 2011.

34. Margaret Livingstone, *Vision and Art: The Biology of Seeing* (New York: Abrams, 2008).

35. Richard G. Stevens, "Artificial Lighting in the Industrialized World, Circadian Rhythms and Breast Cancer," *Cancer Causes and Control* 17 (2006): 501–507.

36. Dennis Dutton, *The Art Instinct: Beauty, Pleasure and Human Evolution* (New York: Bloomsbury Press, 2009), 19–20.

37. Pallasmaa, "Space, Place and Atmosphere."

38. Hull, *Touching the Rock*, 30.

39. Michael Arbib, lecture at "Minding Design."

40. Noë, *Out of Our Heads*, 23–24.

41. Ibid.

42. Louis Sullivan, *Kindergarten Chats* (New York: Dover, 1979), 67.

43. This is the central argument in Robinson, *Nesting*.

44. David Abrams, "The Air Aware: Mind and Mood on a Breathing Planet," *Orion Magazine*, September/October 2009.

45. Mark Johnson, "The Embodied Meaning of Architecture," chapter 2 above in this volume.

46. Johnson, *The Meaning of the Body*, 117.

47. Alberto Pérez-Gómez, *Architecture and the Crisis of Modern Science* (Cambridge, MA: MIT Press, 1984), 12.

48. Richard Neutra, *Survival through Design* (New York: Oxford University Press, 1954), 12.

49. John Dewey, *Art as Experience* (New York: Penguin, 2005), 224.

50. Johan Wolfgang von Goethe, "Allerdings," *Dem Physiker*, 1828.

51. Maurice Merleau-Ponty, *The Merleau-Ponty Reader*, ed. Ted Toadvine and Leonard Lawlor (Chicago: Northwestern University Press, 2007), 56.

8

EMBODIED SIMULATION, AESTHETICS, AND ARCHITECTURE: AN EXPERIMENTAL AESTHETIC APPROACH

Vittorio Gallese and Alessandro Gattara

Every human contact with the things of the world contains both a meaning- and a presence-component. ... The situation of aesthetic experience is specific inasmuch as it allows us to live both these components in their tension.

Hans Gumbrecht[1]

Cognitive neuroscience today offers a novel approach to the study of human social cognition and culture. Such an approach can be viewed as a sort of "cognitive archaeology," as it enables the empirical investigation of the neurophysiological brain mechanisms that make our interactions with the world possible, thereby allowing us to detect the possible functional antecedents of our cognitive skills and to measure the sociocultural influence exerted through human cultural evolution on that very same cognitive repertoire. Thanks to cognitive neuroscience we can deconstruct some of the concepts we normally use when referring to intersubjectivity or to aesthetics, art, and architecture, as well as when considering our experience of them.

This chapter, written by a cognitive neuroscientist and an architect, endeavors to suggest why and how cognitive neuroscience should investigate our relationship with aesthetics and architecture—framing this empirical approach as experimental aesthetics. The term experimental aesthetics specifically refers to the scientific investigation of the brain-body physiological correlates of the aesthetic experience of particular human symbolic expressions, such as works of art and architecture. The notion "aesthetics" is used here mainly in its bodily connotation, as it refers to the sensorimotor and affective aspects of our experience of these particular perceptual objects.

Of course, this approach covers only one aspect of aesthetics, since it refers to an early component of our perceptual experience of the object: to what is happening before any explicit judgment is formulated. The neurophysiological and behavioral evidence of this early phase of aesthetic experience is strikingly similar to that which underlies the mundane perceptual experience of nonartistic objects. Thus, experimental aesthetics can also clarify how different the neurophysiological and bodily correlates of "real-world" experience are from those that characterize experiencing the symbolic representations of that world. We address some recently discovered multimodal properties of the motor system, introducing mirror neurons and embodied simulation, and discuss their relevance for an embodied account of aesthetic experience, summarizing recent empirical research that targets the relationship between gestures and meaning-making. We conclude by proposing several suggestions on how experimental aesthetics might help us to understand the experience of architecture. We believe that only a multidisciplinary approach can increase our understanding of these important and distinctive aspects of human culture.

FOUR REASONS WHY COGNITIVE NEUROSCIENCE MATTERS TO ARCHITECTURE

Cognitive neuroscience is not an alternative to the humanities but a different methodological approach that explains the same phenomena with a different epistemological attitude, a different level of description, and a different language. Cognitive neuroscience can contribute to addressing the following questions: What does it mean "to look at" a painting, a Greek temple or a film, in terms of the brain-body system? To what extent does the way we experience "reality" and fiction depend upon different epistemic approaches and different underpinning neurofunctional mechanisms?

The reasons why cognitive neuroscience is entitled to formulate such questions, and supposedly also to help in answering them, are the following, listed according to their decreasingly broad implications: The first reason deals with the relationship between perception and empathy. For many years aesthetics and cognitive science have shared a particular attitude toward the sense of vision when accounting for aesthetic experience and

8.1 Greek temple at Segesta, Sicily.

the perceptual representation of the world, respectively. Both approaches endorsed a sort of "visual imperialism," neglecting the multimodal nature of vision. In the following section we demonstrate that such a notion of vision no longer holds, and introduce neuroscientific evidence of the relationship between the motor system, the body, and the perception of space, objects, and the actions of others.

The notion of empathy, recently explored by cognitive neuroscience, can reframe the problem of how art functions and how architectural spaces are experienced—revitalizing and empirically validating old intuitions about the relationship between body, empathy, and aesthetic experience.

The second reason addresses how the real and fictional worlds relate to one another and to the brain-body system. Empirical research has shown that we experience fictional realities through neurobiological mechanisms fairly similar to those through which we experience real life. We show how, from a certain point of view, any experience of any possible world basically depends upon similar embodied simulation routines. The "as-if" mode of embodied simulation appears to qualify not just our appreciation of fictional worlds, but all forms of intentional relations, including those characterizing our prosaic daily reality.

The third reason deals with architecture and its aesthetic quality. Embodied simulation can illuminate the aesthetic aspects of architecture—both from the perspective of its making, as well as the potential experience it affords the beholder—by revealing the intimate intersubjective nature of any creative act: where the physical object, the product of symbolic expression, becomes the mediator of an intersubjective relationship between creator and beholder. The experience of architecture, from the contemplation of the decorative element of a Greek temple to the physical experience of living and working within a specific architectonic space, can be deconstructed into its grounding bodily elements. Cognitive neuroscience can investigate of what the sense of presence that some buildings possess is made. This approach can also contribute a fresher empirical take on the evolution of architectonic style and its cultural diversity, by treating it as a particular case of symbolic expression, and through identifying its bodily roots.

THE MULTIMODAL NATURE OF VISION

Observing the world is more complex than the mere activation of the visual brain. Vision is multimodal; it encompasses the activation of motor, somatosensory, and emotion-related brain networks. Any intentional relation we might entertain with the external world has an intrinsic pragmatic nature; hence it always bears a motor content. More than five decades of research have shown that motor neurons also respond to visual, tactile, and auditory stimuli. The same motor circuits that control the motor behavior of individuals also map the space around them, the objects at hand in that very same space, thus defining and shaping in motor terms their representational content.[2] The space around us is defined by the motor potentialities of our body. Premotor neurons controlling the movements of the upper arm also respond to tactile stimuli applied to it, to visual stimuli moved within the arm's peripersonal space, or to auditory stimuli also coming from that same peripersonal space.[3] The manipulable objects we look at are classified by the motor brain as potential targets of the interactions we might entertain with them. Premotor and parietal "canonical neurons" control the grasping and manipulation of objects and respond to their mere observation, as well.[4] Finally, mirror neurons—motor neurons activated during the execution of an action and its observation performed by someone else—map the action of others on the observer's motor representation of the same action.[5]

More than twenty years of research on mirror neurons have demonstrated the existence of a mechanism directly mapping action perception and execution in the human brain, here defined as the mirror mechanism (MM).[6] Also, in humans, the motor brain is multimodal. Thus, it does not matter whether we see or hear the noise made by someone cracking peanuts, or locking a door. Different—visual and auditory—sensory accounts of the same motor behavior activate the very motor neurons that normally enable the original action. The brain circuits showing evidence of the MM, connecting frontal and posterior parietal multimodal motor neurons, most likely analogous to macaques' mirror neurons, map a given motor content like "reach out" or "grasp" not only when controlling its performance, but also when perceiving the same motor behavior performed by someone else, when imitating it, or when imagining performing it while being perfectly still.

These results completely change our understanding of the role of the cortical motor system and of bodily actions. The cortical motor system is not just a movement machine, but an integral part of our cognitive system,[7] because its neurofunctional architecture structures not only action execution but also action perception, imitation, and imagination, with neural connections to motor effectors and/or other sensory cortical areas. When the action is executed or imitated, the corticospinal pathway is activated, leading to the excitation of muscles and the ensuing movements. When the action is observed or imagined, its actual execution is inhibited. The cortical motor network, though, is not activated in all of its components and not with the same intensity, hence action is not produced—it is only simulated.

The prolonged activation of the neural representation of motor content in the absence of movement probably defines the experiential backbone of what we perceive or imagine perceiving. This allows a direct apprehension of the relational quality linking space, objects, and others' actions to our body. The primordial quality turning space, objects, and behavior into intentional objects is their constitution as objects of the motor intentionality that our body's motor potentialities express.[8]

Other MMs seem to be involved with our capacity to directly apprehend the emotions and sensations of others due to a shared representational bodily format. When perceiving others expressing disgust, or experiencing touch or pain, the same brain areas are activated as when we subjectively experience the same emotion or sensation. We do not fully experience their qualitative content, which remains opaque to us, but its simulation instantiated by the MM enables us to experience the other as experiencing emotions or sensations we know from the inside, as it were.

EMBODIED SIMULATION AND THE EMPATHIC BODY

The discovery of mirror neurons provides a new, empirically founded notion of intersubjectivity connoted first and foremost as intercorporeality—the mutual resonance of intentionally meaningful sensorimotor behaviors. Our understanding of others as intentional agents does not *exclusively* depend on propositional competence, but also on the relational nature of action. In many situations we can directly understand the meaning of other people's basic actions thanks to the motor equivalence between what others do and what we *can* do. Intercorporeality thus becomes the main source of knowledge we have of others. Motor simulation instantiated by neurons endowed with "mirror properties" is probably the neural correlate of this human faculty, describable in functional terms as "embodied simulation."[9]

The multiple MMs present in our brain, thanks to the "intentional attunement" they generate, allow us to recognize others as other selves, allowing basic forms of intersubjective communication and mutual implicit understanding.[10] Embodied simulation provides a unified theoretical framework for all of these phenomena. It proposes that our social interactions become meaningful by means of reusing our own mental states or processes in functionally attributing them to others. In this context, simulation is conceived as a nonconscious, prereflective functional mechanism of the brain-body system, whose function is to model objects, agents, and events. This mechanism can be triggered during our interactions with others, since it is being plastically modulated by contextual, cognitive, and personal identity-related factors.

Embodied simulation is also triggered during the experience of spatiality around our body and during the contemplation of objects. The functional architecture of embodied simulation seems to constitute a basic characteristic of our brain, making possible our rich and diversified experiences of space, objects, and other individuals, and is the basis of our capacity to empathize with them.

Taken together, the results summarized thus far suggest that empathy—or, at the very least, many of its bodily qualities—might be underpinned by embodied simulation mechanisms. According to our proposal, empathy is the outcome of the natural tendency to experience our interpersonal relations fundamentally at the implicit level of intercorporeality: that is, at the level of the mutual resonance of intentionally meaningful sensory-motor behaviors.

It is perhaps worth emphasizing that embodied simulation not only connects us to others, it connects us to *our* world—a world populated by natural and man-made objects, with or without a symbolic nature, and with other individuals: a world in which, most of the time, we feel at home. The sense we attribute to *our* lived experience of the world is grounded in the affect-laden relational quality of *our* body's action potentialities, enabled by the way they are mapped in *our* brains.

EMPATHY, EMBODIED SIMULATION, AND AESTHETIC EXPERIENCE

The idea that the body might play an important role in the aesthetic experience of visual art is quite old. The notion of empathy (*Einfühlung*) was originally introduced to aesthetics in 1873 by the German philosopher Robert Vischer, well before its use in psychology. Vischer described *Einfühlung*, literally "feeling-in," as the physical response generated by the observation of forms within paintings. Particular visual forms arouse particular responsive feelings, depending on the conformity of those forms to the design and function of the muscles in the body, from our eyes to our limbs and to our bodily posture as a whole. Vischer clearly distinguished a passive notion of vision—*seeing*—from the active one of *looking*. According to Vischer, *looking* best characterizes our aesthetic experience when perceiving images, in general, and works of art, in particular.

Aesthetic experience implies an empathic involvement encompassing a series of bodily reactions of the beholder. In his book *On the Optical Sense of Form*, Vischer wrote: "We can often observe in ourselves the curious fact that a visual stimulus is experienced not so much with our eyes as with a different sense in another part of our body. ... The whole body is involved; the entire physical being [*Leibmensch*] is moved. ... Thus each emphatic sensation ultimately leads to a strengthening or a weakening of the general vital sensation [*allgemeine Vitalempfindung*]."[11]

Vischer posits that symbolic forms acquire their meaning predominantly because of their intrinsic anthropomorphic content. Through the nonconscious projection of her/his body, the beholder establishes an intimate relation with the artwork.

Developing Vischer's ideas further, Heinrich Wölfflin speculated on the ways in which observation of specific architectural forms engages the beholder's bodily responses.[12] Shortly afterward, Theodor Lipps discussed the relationship between space and geometry on the one hand, and aesthetic enjoyment on the other.[13]

The work of Vischer exerted an important influence over two other German scholars whose contributions are highly relevant for our proposal: Adolf von Hildebrand and Aby Warburg. In 1893, the German sculptor Hildebrand published a book entitled *The Problem of Form in Figurative Art*. In this book Hildebrand proposed that our perception of the spatial characters of images is the result of a constructive sensory-motor process. Space, according to Hildebrand, does not constitute an a priori of experience, as suggested by Kant, but is itself a product of experience. That is to say, artistic images are effective because they are the outcome of both the artist's creative production and the effect that the images elicit in the beholder. The aesthetic value of works of art resides in their potential to establish a link between the intentional creative acts of the artist and the reconstruction of those acts by the beholder. In this way, creation and artistic fruition are directly related. To understand an artistic image, according to Hildebrand, means implicitly grasping its creative process.

A further interesting and very modern aspect of Hildebrand's proposal concerns the relevance he assigns to the motor nature of experience. Through movement, the available elements in space can be connected; objects can be carved out of their background and perceived as such. Through movement, representations and meaning can be formed and articulated. Ultimately, according to Hildebrand, sensible experience is possible, and images acquire their meaning only because of the acting body.

Hildebrand strongly influenced another famous German scholar, Aby Warburg. From 1888 to 1889 Warburg studied in Florence at the Kunsthistorisches Institut, founded by the art historian August Schmarsow. As Didi-Huberman emphasizes, Schmarsow (1853–1936) was determined to open art history to the contributions of anthropology, physiology, and psychology, and emphasized the role of body gestures in visual art, arguing that bodily empathy greatly contributes to the appreciation of visual arts.[14] As Andrea Pinotti writes, Schmarsow, "art historian and theoretician, centered his reflections, which exploited both the results of the theories of empathy and the analyses of the formal character of art works, on the idea of the transcendental function of corporeality as a constellation of material a priori, that is, on the idea of bodily organization as the condition of the possibility of sensory experience."[15]

Warburg clearly learned this lesson, as he conceived art history as a tool that would enable a deeper understanding of the psychology of human expressive power. His famous notion of a "pathemic form" (*Pathosformel*) of expression implies that a variety of bodily postures, gestures, and actions can be constantly detected in art history, from classical art

8.2 Laocoön and his sons. Vatican Museum, Rome.

to the Renaissance period, just because they embody, in an exemplary fashion, the aesthetic act of empathy as one of the main creative sources of artistic style. According to Warburg, a theory of artistic style must be conceived as a "pragmatic science of expression" (*pragmatische Ausdruckskunde*).

Warburg, writing about the classic marble group known as the *Laocoön*, identified transition as a fundamental element that turns a static image into one charged with movement and pathos. Years later, the Russian movie director Sergei Eisenstein, commenting on the same *Laocoön* sculpture in 1935, wrote that the lived expression of human sufferance portrayed in this masterpiece of classical art is accomplished by means of the illusion of movement.[16] Movement illusion is accomplished by means of a particular

8.3 Marcel Duchamp, *Nude Descending a Staircase (No. 2)*, 1912. Oil on canvas. The Philadelphia Museum of Art.

montage, condensing in one single image different aspects of expressive bodily movements that could not possibly be visible at the same time. A similar effect can be appreciated in Duchamp's *Nude Descending a Staircase*.

Maurice Merleau-Ponty further highlighted the relationship between embodiment and aesthetic experience by suggesting the relevance for art appreciation of the felt bodily imitation of what is seen in the artwork.[17] Consistent with the role of *Einfühlung*, Merleau-Ponty also emphasized the importance of the artist's implied actions for the aesthetic experience of the beholder, taking as his example the paintings of Cézanne, when he famously stated that we cannot possibly imagine how a mind could paint.[18]

These scholars believed that the feeling of physical involvement with a painting, a sculpture, or an architectural form provokes a sense of imitating the motion or action seen or implied in the work, while enhancing our emotional responses to it. Thus, physical involvement constitutes a fundamental ingredient of our aesthetic experience of artworks. In the next section we discuss recent empirical evidence confirming bodily empathy as an important component of the perceptual experience of works of art, demonstrating its underlying neural mechanisms.

EMBODIED SIMULATION AND EXPERIMENTAL AESTHETICS

Embodied simulation can be relevant to aesthetic experience in at least two ways. First, because we relate to the bodily feelings triggered by works of art by means of the MMs they evoke. In this way, embodied simulation generates the peculiar "seeing-as" that characterizes our aesthetic experience of the images we look at. Second, the potentially intimate relationship between the symbol-making gesture and its eventual reception by beholders is enabled through the motor representation that produced the image by means of simulation.[19] When I look at a graphic sign, I unconsciously simulate the gesture that produced it.

Our scientific investigation of experimental aesthetics applied to visual arts began with this second aspect. In three distinct experiments, we investigated, by means of high-density electroencephalography (EEG), the link between the expressive gestures of the hand and the images produced by those gestures. We recorded beholders' brain responses to graphic signs like letters, ideograms, and scribbles, or to abstract art works by Lucio Fontana and Franz Kline.

The results of the first study showed that observing a letter of the Roman alphabet, a Chinese ideogram, or a meaningless scribble, all written by hand, activated the hand motor representation of beholders.[20] In the two other studies we demonstrated that a similar motor simulation of hand gestures is evoked when beholding a cut on canvas by Lucio Fontana,[21] or the dynamic brushstrokes on canvas by Franz Kline.[22]

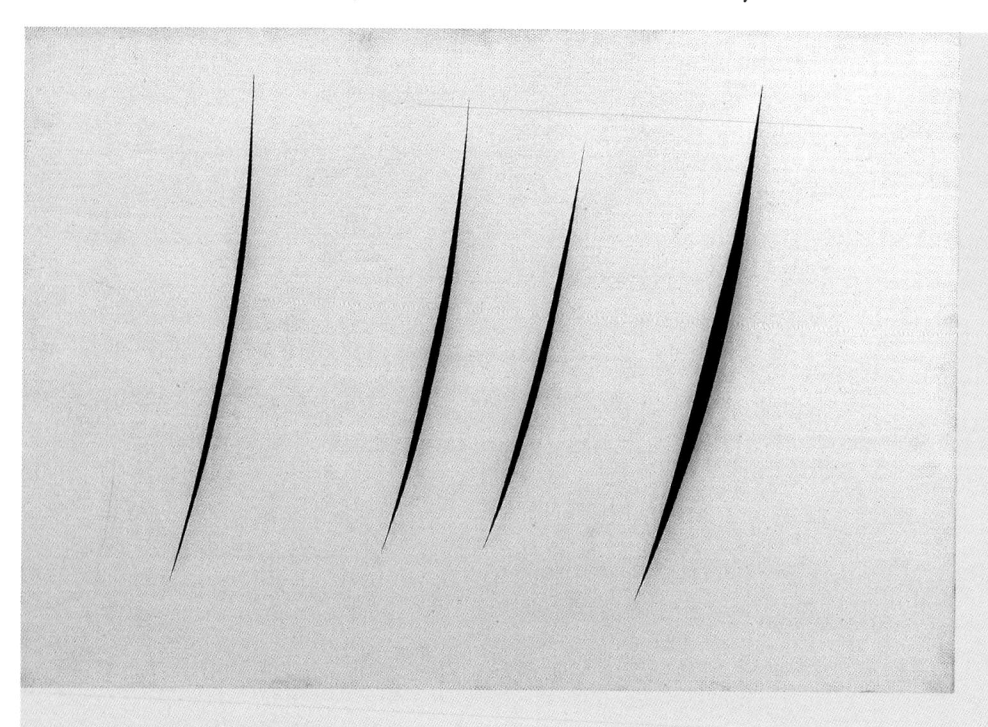

8.4 Lucio Fontana, *Spatial Concept #2*, 1960. Oil on canvas. The Philadelphia Museum of Art.

The visible traces of the creative gesture activated the specific motor areas that control the execution of that same gesture in the observer. Beholders' eyes not only capture information about the shape, direction, and texture of the cuts or strokes, by means of embodied simulation, they emulate the actual motor expression the artist used when creating the artwork. The sensory-motor component of our image perception, together with the jointly evoked emotional reaction, allows beholders to feel the artwork in an embodied manner.

A possible criticism of this model might point to the apparent passivity of this account of aesthetic experience, in which beholders are relegated to a deterministic empathic receptivity, hence losing sight of the peculiar individual quality of aesthetic experience that is largely determined by one's individual taste, background, memories, education, and expertise.

8.5 Franz Kline, *Chief*, 1950. Oil on canvas. The Museum of Modern Art, New York.

A second objection frequently raised against empathic-mimetic accounts of aesthetic experience consists of opposing the ambiguity and indeterminacy of art's symbolic content to the supposedly mechanistic quality of empathic responses, hence falling short of capturing the potential intrinsic ambiguity and polysemic quality of works of art.

We think it is possible to counter these criticisms by arguing that there is ample evidence that MMs and embodied simulation are dynamically modulated, and affected by contingent and idiosyncratic factors. Several studies have shown that one's previous experiences, memories, and expertise strongly determine the intensity of activation of MMs and the ensuing perceptual contents.[23]

We posit that embodied simulation, by virtue of its diachronic plasticity and modulation, might be also the vehicle for the projective qualities of our aesthetic experience—where our personal and social identity literally shape the way we relate to a given perceptual object. Embodied simulation, if conceived of as the dynamic instantiation of our implicit memories, can relate the perceptual object and beholder to a specific, unique, and historically determined quality. This projective quality of embodied simulation refutes both objections.

EXPERIMENTAL AESTHETICS AND ARCHITECTURE: SUGGESTIONS FOR A ROAD MAP

We have already referred to Heinrich Wölfflin as one of the earliest proponents of the relationship between our bodily nature and our experience of architecture. According to Wölfflin, if we were merely visual creatures, the aesthetic appreciation of works of art and architecture would be precluded. The very nature of our body allows us to experience gravity, force, and pressure, and thus makes the enjoyment of contemplating a Doric temple, or the feeling of being elevated when entering a Gothic cathedral, possible in the first place. Furthermore, we offered a concise account of why the available empirical neuroscientific evidence seems to support this view.

We can now empirically test this view by recording the brain and bodily responses of volunteers perceptually experiencing and exploring virtual architectonic environments by means of immersive virtual reality. Today, virtual caves can reproduce high-accuracy, three-dimensional and richly dimensioned digital versions of temples, squares, churches, and buildings in which individuals can not only enjoy a vivid and realistic experience but may also virtually explore as if they are moving around, directing their gaze at different details and spatial locations. The ecological plausibility of such virtual experience can be established in the absence of any active movement on the part of beholders, making these ideal conditions in which to record brain signals and autonomic bodily responses, thereby minimizing movement-driven artifacts and signal noise.

This experimental approach could enable us to empirically address important aspects of architectural history, like the evolution of architectonic style, by charting its potential biological bodily roots. The same approach could also shed light on the plausibility of hypotheses about the supposed biomorphic and /or anthropomorphic origin of architectonic elements and decorations.[24] A second possible application of this approach to architecture deals with the relationship between architectural spaces and the way they are experienced by the people living and working in them.

Juhani Pallasmaa has criticized Western culture's excessive "oculocentrism," the overriding tendency that assigns a cognitive privilege to vision. With the invention of perspective, the eye becomes at once the center of the perceptual world and the center of the subject perceiving that world.[25] According to Pallasmaa, the scopic regime instantiated by visual perspective exemplifies the disembodied nature of the Cartesian subject, whose solipsism segregates the mind from the body, the subject from the object, and the I from the Thou. Such a "purovisibilist" attitude deeply influenced contemporary architecture by predominantly adhering, according to Pallasmaa, to a purely formalist perspective; as a consequence it has lost contact with the very people for whom the architectural project was originally intended.

As Sarah Robinson comprehensively reviews the body schema earlier in this book, she writes: "peripersonal space describes the space immediately surrounding our bodies; extrapersonal space refers to the space just beyond the peripersonal."[26] The constant weighting of architectonic and peripersonal space is mainly processed by premotor neurons mapping visual space on potential action or motor schemata. Furthermore, MMs for action are modulated by proxemics, as the potentiality for interactions between agent and observer measured by the distance separating them can affect the intensity of the discharge of mirror neurons in the beholder's brain.[27]

As the experience of the built environment and its affordances is shaped through the precognitive activation of motor simulations, the role of embodied simulation in architectural experience becomes even more interesting if one considers emotions and sensations. A typical and recurring experience in everyday life is reacting with positive or negative feelings upon opening a door and going, for the first time, into a new architectural environment. Though, as Harry Mallgrave notes, to date little, if any, neuroimaging research has been done on the emotional experience of architectural environments.[28] The same applies to the haptic qualities of materials employed to design exterior and interior parts of architectonic spaces whose multimodal impact and desirability could be easily measured.

The knowledge acquired through experimental aesthetics might provide new insights, just to mention the most obvious ones, for the future design of office spaces or retail stores. Both are usually designed with strict and short deadlines by architectural firms specialized and routinized in this building typology. Such firms frequently need new fit-outs and refurbishments to stay ahead of challenging competitors.

The open office was originally conceived in the 1950s in Germany as *Bürolandschaft*, or office landscape, to facilitate communication and idea flow, and has since seen a dramatic growth in use (70 percent of all offices now have an open floor plan) as well as increasing levels of frustration in the employees who work in such environments. While proxemics contributed to more effective distribution of employees in open office layouts,

8.6 Office interior of Lombardini 22, DEGW Milano.

a neuroscientific approach to the study of peripersonal space could help architects shape working environments that ultimately promote employees' well-being and productivity.

As recently shown by architect Isabella Pasqualini, neuroscientist Olaf Blanke, and colleagues, "investigating bodily feelings, self-identification and self-location with respect to the architectonic unity or form and space of the observer may be confronted with architecture as an extension of the bodily volume."[29] According to this study, the experience of a narrow room increases the somatosensory sensation of verticality, "thus enhancing bodily stability," while conversely, a large room elicits "a destabilizing effect for the missing cue in peripersonal space provoking an illusory backwards movement."[30]

It is interesting to emphasize that these results prove to be coherent with Schmarsow's notion of space "from within." According to Schmarsow, "every spatial creation is first and foremost the enclosing of a subject."[31] Indeed, the motor system is also responsible for the phenomenal awareness of the body's relations with the environment. We are planning to study how daily actions or social interactions virtually presented within differently designed architectonic spaces are experienced differently by beholders. We will also study if and how such different experiences correlate with different profiles of bodily and brain responses.

CONCLUSIONS

Even if the notions of embodiment and empathy within the architectural field are much older than cognitive neuroscience itself, the latter is shedding new light on a topic that is otherwise dismissed or neglected by mainstream theory. The theory of empathy began to have an impact within the contemporary architectural field—as Harry Mallgrave insightfully described—in the "Garden City" of Hellerau, an experiment by Wolf Dohrn on the outskirts of Dresden in 1908–1914; in the Bauhaus in 1919–1933; and in Richard Neutra's book *Survival through Design* in 1954. Architects and architectural scholars such as Juhani Pallasmaa, Steven Holl, Alberto Pérez-Gómez, and Harry Mallgrave have revitalized and brought the topic of empathy back into contemporary discourse, some of them in this volume.

Architecture is among the fruits of the new way in which humans, at a given point in their cultural evolution, were able to relate to the external world. The material world was no longer exclusively considered to be a domain to exploit for the utilitarian satisfaction of biological needs. Material objects lost their unique status as tools to become symbols, public epiphanies able to make something that is absent visible—to make tangible something that apparently is present only in the mind of the creator and the beholder. Humans, thanks to the expression of their symbolic creativity, acquired the capacity to give shape to material objects, conferring on them the meaning they intrinsically lack. Such meaning is the outcome of the creator's action of collectively building a temple or a cathedral, laying colors on a canvas, or turning a marble block into a *David* or a *Rape of Proserpina*.

Today, cognitive neuroscience can reveal—from its own peculiar perspective and methodology—the aesthetic quality of human nature and our natural creative inclination. This new research will help us to understand how and why art and architecture are among the most fundamental expressions of our human nature.

NOTES

Many of the ideas and proposals presented in this chapter were published in a slightly different form in these recent papers: V. Gallese and C. Di Dio, "Neuroesthetics: The Body in Aesthetic Experience," in V. Ramachandran et al., *Encyclopedia of Human Behavior*, 2nd edn. (Amsterdam: Elsevier, 2012), vol. 2, 687–693; V. Gallese, "Bodily Selves in Relation: Embodied Simulation as Second-Person Perspective on Intersubjectivity," *Philosophical Transactions of the Royal Society B* 369 (2014): 20130177; V. Gallese, "The Hand and the Architect: Gesture and Creative Expression," in *Unplugged: Projects of L22 and DGW Italy* (Milan: L22, 2014), 14–17; V. Gallese, "Arte, corpo, cervello: Per un'estetica sperimentale," *Micromega* 2 (2014): 49–67.

1. Hans Ulrich Gumbrecht, *Production of Presence: What Meaning Cannot Convey* (Stanford: Stanford University Press, 2004), 109.

2. V. Gallese, "The Inner Sense of Action: Agency and Motor Representations," *Journal of Consciousness Studies* 7 (2000): 23–40; G. Rizzolatti, L. Fogassi, and V. Gallese, "Motor and Cognitive Functions of the Ventral Premotor Cortex," *Current Opinion in Neurobiology* 12 (2002): 149–154.

3. L. Fogassi, V. Gallese, L. Fadiga, G. Luppino, M. Matelli, and G. Rizzolatti, "Coding of Peripersonal Space in Inferior Premotor Cortex (Area F4)," *Journal of Neurophysiology* 76 (1996): 141–157; G. Rizzolatti, L. Fadiga, L. Fogassi, and V. Gallese, "The Space around Us," *Science* 277 (1997): 190–191.

4. A. Murata, L. Fadiga, L. Fogassi, V. Gallese, V. Raos, and G. Rizzolatti, "Object Representation in the Ventral Premotor Cortex (Area F5) of the Monkey," *Journal of Neurophysiology* 78 (1997): 2226–2230; V. Raos, M. A. Umiltà, L. Fogassi, and V. Gallese, "Functional Properties of Grasping-Related Neurons in the Ventral Premotor Area F5 of the Macaque Monkey," *Journal of Neurophysiology* 95 (2006): 709–729.

5. G. di Pellegrino, L. Fadiga, L. Fogassi, V. Gallese, and G. Rizzolatti, "Understanding Motor Events: A Neurophysiological Study," *Experimental Brain Research* 91 (1992): 176–180; Fogassi, Gallese, Fadiga, et al., "Coding of Peripersonal Space in Inferior Premotor Cortex"; G. Rizzolatti, L. Fadiga, V. Gallese, and L. Fogassi, "Premotor Cortex and the Recognition of Motor Actions," *Cognitive Brain Research* 3 (1996): 131–141; G. Rizzolatti, L. Fogassi, and V. Gallese, "Neurophysiological Mechanisms Underlying the Understanding and Imitation of Action," *Nature Reviews Neuroscience* 2 (2001): 661–670.

6. M. Ammaniti and V. Gallese, *The Birth of Intersubjectivity: Psychodynamics, Neurobiology and the Self* (New York: W. W. Norton, 2014), 236; Gallese, "Bodily Selves in Relation."

7. V. Gallese, M. Rochat, G. Cossu, and C. Sinigaglia, "Motor Cognition and Its Role in the Phylogeny and Ontogeny of Intentional Understanding," *Developmental Psychology* 45 (2009): 103–113.

8. Gallese, "The Inner Sense of Action"; V. Gallese and C. Sinigaglia, "What Is So Special with Embodied Simulation," *Trends in Cognitive Sciences* 15 (2011): 512–519; Gallese, "Bodily Selves in Relation."

9. Gallese, "Bodily Selves in Relation"; Gallese and Sinigaglia, "What Is So Special with Embodied Simulation."

10. Gallese, "Bodily Selves in Relation."

11. Robert Vischer, *Über das optische Formgefühl: Ein Beiträg zur Ästhetik* (Leipzig: Credner, 1872), 98–99.

12. Heinrich Wölfflin, *Prolegomena zu einer Psychologie der Architektur* (Berlin, 1886).

13. Theodor Lipps, "Einfühlung, innere nachahmung und organenempfindung," *Archiv für die gesamte Psychologie* 1 (1903): 185–204.

14. Georges Didi-Huberman, *L'immagine insepolta* (Milan: Bollati Boringhieri, 2006).

15. Andrea Pinotti, *Memorie del neutro. Morfologia dell'immagine in Aby Warburg* (Milan: Mimesis, 2001), 91.

16. Sergei Eisenstein, *Towards a Theory of Montage*, trans. Michael Glenny, vol. 2 of Eisenstein, *Selected Works* (1935; London: I. B. Tauris, 2010).

17. Maurice Merleau-Ponty, *Phenomenology of Perception*, trans. C. Smith (London: Routledge, 1962); Maurice Merleau-Ponty, *The Visible and the Invisible*, trans. A. Lingis (Evanston: Northwestern University Press, 1968).

18. Ibid.

19. D. Freedberg and V. Gallese, "Motion, Emotion and Empathy in Aesthetic Experience," *Trends in Cognitive Sciences* 11 (2007): 197–203; see also Gallese and Di Dio, "Neuroesthetics"; Gallese, "Bodily Selves in Relation"; Gallese, "The Hand and the Architect"; Gallese, "Arte, corpo, cervello."

20. K. Heimann, M. A. Umiltà, and V. Gallese, "How the Motor-Cortex Distinguishes among Letters, Unknown Symbols and Scribbles: A High Density EEG Study," *Neuropsychologia* 51 (2013): 2833–2840.

21. M. A. Umiltà, C. Berchio, M. Sestito, D. Freedberg, and V. Gallese, "Abstract Art and Cortical Motor Activation: An EEG Study," *Frontiers in Human Neuroscience* 6 (2012): 311.

22. B. Sbriscia-Fioretti, C. Berchio, D. Freedberg, V. Gallese, and M. A. Umiltà, "ERP Modulation during Observation of Abstract Paintings by Franz Kline," *PLoS ONE* 8 (2013): e75241.

23. For a recent review, see Gallese, "Bodily Selves in Relation."

24. R. B. Onians, *The Origins of European Thought about the Body, the Mind, the Soul, the World, Time and Fate* (Cambridge: Cambridge University Press, 1951); Vincent Scully, *The Earth, the Temple and the Gods* (New Haven: Yale University Press, 1962); Joseph Rykwert, *On Adam's House in Paradise: The Idea of Primitive Hut in Architectural History* (New York: Museum of Modern Art, 1972); Sarah Robinson, *Nesting: Body, Dwelling, Mind* (Richmond, CA: William Stout, 2011); Harry Francis Mallgrave, *Architecture and Embodiment: The Implications of the New Sciences and Humanities for Design* (New York: Routledge, 2013).

25. Juhani Pallasmaa, *The Eyes of the Skin: Architecture and the Senses*, 2nd edn. (Chichester, UK: John Wiley, 2005), 16.

26. Robinson, "Nested Bodies," chapter 7 above in this volume.

27. For a review, see Ammaniti and Gallese, *The Birth of Intersubjectivity*.

28. Harry Francis Mallgrave, "'Know Thyself,'" chapter 1 above in this volume.

29. I. Pasqualini, J. Llobera, and O. Blanke, "'Seeing' and 'Feeling' Architecture: How Bodily Self-Consciousness Alters Architectonic Experience and Affects the Perception of Interiors," *Frontiers in Psychology* 4 (2013): art. 354.

30. Ibid.

31. August Schmarsow, *Das Wesen des architektonischen Schöpfung* (Leipzig: Karl W. Hiersemann, 1894).

9

FROM INTUITION TO IMMERSION: ARCHITECTURE AND NEUROSCIENCE

Melissa Farling

If you enter the Cathedral in Amiens at twilight while an organ is playing and find that your "heart skips a beat," it's because your brain—not your heart—has filled you with awe. Cells in your brain are gorging themselves with a sudden flush of blood, raising your temperature, quickening your pulse, and flooding you with memories. Light flooding through stained glass windows is stimulating the V4 area of your visual cortex. Bach's music is vibrating within the cochlea of your inner ear and sending signals to the auditory cortex. The musty smells of centuries past register unconsciously on the olfactory neurons at the bridge of your nose. You are experiencing architecture.[1]

John P. Eberhard

Eberhard's description of experiencing Amiens Cathedral explains all that a person takes in with her senses, and allows us to understand not only the psychological, but also the neurological and physiological aspects of that experience. This rich, multidimensional understanding was exactly what I was hoping to explore twenty-seven years ago in architecture school.

When I was a third-year architectural student, my studio conducted a charette for each individual student. The goal of the charette was to generate ideas for projects that would combine two passions—one passion had to be architecture, the other was personal to each student. My other passion was psychology; so, at the end of my charette, we determined that I should explore the impacts of architectural spaces on people's behavior. We also decided the best way to pursue this goal would be to limit my research to a building type that already severely restricted a person to the built environment. We narrowed a broad list down to three building types: a monastery, a convent, and a prison. I was immediately drawn to investigating prison design ... the most restrictive of the three environments. A prison is a microcosm of a community, and has an innate mission to "correct" or rehabilitate a person. And there was already an overwhelming amount of environmental psychology and sociological research in which I could immerse myself. This prison project marked the beginning of a deeply passionate journey for me to try to understand and quantify the profound impacts of the built environment.

CREATING THE LINK

Fast forward eighteen years to the 2005 National AIA Convention, where John Paul Eberhard presented a paper as part of a Latrobe Fellowship on the Academy of Neuroscience for Architecture (ANFA). An architect, author, and founding president of ANFA, John had been immersing himself in neuroscience for a decade. John discusses the genesis of ANFA in his chapter "Architecture and Neuroscience: A Double Helix," in this volume. His presentation at the convention, and my discovery that ANFA even existed, overwhelmed me. Here was an organization of leading architects and scientists based in San Diego—with beginnings rooted in the Salk Institute, an architectural icon. ANFA's mission is "to promote and advance knowledge that links neuroscience research to a growing understanding of human responses to the built environment."[2]

Following John's presentation, I introduced myself and ultimately became a Research Associate for ANFA. ANFA board members appointed associates who are neuroscientists, architects, and architectural graduates who worked for one or more years in neuroscience labs or architectural firms to gain knowledge of the new field. It was also an opportunity for associates to spread the word about their experiences and the knowledge they acquired. These goals were accomplished by teaching, conducting new research, publishing the research, participating in workshops, and presenting research to many professional organizations including the AIA.[3]

My guiding words were summarized by John Eberhard at his AIA presentation. The original quote was from Dr. Fred Gage's theme presentation at the 2003 AIA National Convention in San Diego, when ANFA was formed and officially announced:

> while the brain controls our behavior and genes control the blueprint for the design and structure of the brain, the environment can modulate the function of genes and, ultimately, the structure of our brain. Changes in the environment change the brain, and therefore they change our behavior. In planning the environments in which we live, *architectural design changes our brain and our behavior*.[4]

This statement is a compelling reinforcement of the theory that understanding our brains and behavior is essential to creating sensitive and appropriate environments for people. The exploration of neuroscience and architecture has the potential to take an architect's intuition to another level.

DEFINING OUR RESPONSIBILITY

Originally, I entitled this essay "From Intuition to Evidence" until I had a conversation with Max Underwood, President's Professor at Arizona State University Herberger Institute for Design and the Arts. After some discussion, he suggested changing the title to "From Intuition to Immersion." This title is more appropriate on two levels: first, it focuses on the responsibility of architects and our need to immerse ourselves in knowing as much as we can about issues influencing design (geography, topography, climate, program, politics, the sciences ...) to better quantify and communicate the benefits of responsive design to our clients. Additionally, "immersion" speaks to our natural state in our environments and communities—whether consciously or not, the environment powerfully influences us.

VALUES

As a profession, architects have lost significant public credibility when it comes to the importance of the qualitative aspects of design. In recent years other causes, such as building performance, have taken priority, and quite often the public considers design to be a personal focus on aesthetics, and does not understand the deep and rigorous exploration that takes place between the architects, stakeholders, and users to produce technically successful and aesthetically pleasing outcomes. With many building types—especially public buildings—building performance metrics are at the top of the list. High building performance (in terms of energy usage, water usage, carbon footprint, etc.) is an excellent and necessary goal, but it covers only quantitative outcomes, not necessarily qualitative outcomes (although when paired with thoughtful design, the quantitative and qualitative effects can be exponential and very powerful—a compltely separate discussion). As Peter

Buchanan offers in his "Big Rethink" series, sustainability is not just ecology and technology, it is also about psychology and culture.[5] Kenneth Frampton points out in "Towards a Critical Regionalism: Six Points for an Architecture of Resistance" that "the tectonic is not to be confused with the purely technical, for it is more than the simple revelation of stereotomy or the expression of skeletal framework."[6] Psychology and culture are not often measured by architects or clients unless post-occupancy evaluations are conducted. We live in an "age of metrics" focused on measuring building performance, with little value placed on the quality of the individual experience. Building performance needs to be balanced with experiential outcomes, and neuroscience can provide the necessary evidence for architects' intuitive strategies.

Environmental psychologists have conducted the most well-known research linking behavior and architecture. The simple difference between the information that environmental psychology provides and the information neuroscience can offer is this: the environmental psychologists tell us *what* behavior is occurring; the neuroscientist tells us *why* the behavior is occurring. Neuroscience hypotheses are often based on the findings of environmental psychology and/or input from experienced architects, designers, and users. Although intuition is embedded in experience and observation, it is without proof. Fully engaging and understanding all forces impacting architectural experience is an architect's responsibility. The medical profession has the Hippocratic Oath protecting the sick from harm and injustice. The architectural profession should similarly better understand the effects of the buildings we design in order to achieve similar goals.

One neuroscience and architecture team, led by Dr. Eve Edelstein and Dr. Peter Otto, identified at least some of the reasons for millions of lives lost to medical error in hospitals. At the ANFA conference in 2012, they presented compelling evidence demonstrating the various audio problems contributing to these errors. Part of the presentation illustrated the difficulty in distinguishing names of medications at a nurses' station because of the background noise and the similarity of medication names. Understanding the brain processes occurring when we hear is helping to lead to possible design solutions. Assessing intelligibility, sound masking, source location of the sound, building material selections, and physical room layouts were discussed as potential strategies for prevention of future errors.[7]

The United States continues to design and build solitary confinement cells in prisons even though we know they harm. Modern solitary confinement typically refers to placing an inmate in an isolated cell with few or no amenities, often for 23 hours a day.[8] Two years ago, the United Nations determined that solitary confinement for over 15 days is a form of torture. In the past few years, psychiatrists have given substantial expert testimony to Congress about the impacts of the conditions on inmates' behavior—solitary confinement causes or exacerbates mental illness, leading to suicide. Other variables in addition to the built environment must be considered with solitary imprisonment, but design is a major factor, and understanding how these spaces affect the brain would provide powerful information. Frank Gehry, Pritzker-winning architect, in a discussion in 2006 with Dr. Fred Gage, commented that applying neuroscience to architecture is too "prescriptive." Yet Gehry also spoke of the benefit of "informed intuition."[9] Neuroscience research could add another layer of pertinent information to "informed intuition"—much as a building program or an analysis of site conditions provides additional design criteria.

Historically, architects have sought to define and communicate the ethos of architecture. Vitruvius identified fundamental principles of architecture such as symmetry, harmony, and proportion in his *Ten Books on Architecture.*[10] Christopher Alexander, Sara Ishikawa, and Murray Silverstein wrote *The Timeless Way of Building*, *The Oregon Experiment*, and *A Pattern Language: Towns, Buildings, Construction.* This last work creates a language to empower people to design their environment, and works to define hard-to-quantify issues such as the "magic of the city"; designing safe environments in which children can explore and grow; promoting social interaction at home and in a community; making comfortable connections to neighbors and nature; and creating exciting and cheerful spaces.[11] The authors explore the problems of modernity that Paul Ricoeur framed so well when he wrote: "It is a fact: every culture cannot sustain and absorb the shock of modern civilization. There is the paradox: how to become modern and return to sources; how to revive an old dormant civilization and take part in universal civilization."[12]

In his fifth and sixth points of "Towards a Critical Regionalism," Frampton emphasizes topography, context, climate, light, and tectonic form rather than scenographic (visual) features. He places great importance on our capacity to "read the environment in terms other than those of sight alone."[13] He discusses light, darkness, heat and cold, humidity, aroma, presence, momentum and echo of materials. Frampton focuses on our perceptions and how they impact design. The benefit of linking neuroscience to practice is the ability to immerse ourselves in all aspects and effects of design—by taking a more holistic approach to problem-solving, which includes as its very basis a more thorough understanding of the human experience.

EDUCATION

In Buchanan's "Big Rethink" series, the responsibility of the architectural profession is challenged in an effort to reconsider our approaches to sustainability, urbanism, and education. In one of Buchanan's essays, "Integral Theory," he proposes an approach to architectural education that integrates several fields of study. In a subsequent essay, he explains the need to redefine design and creativity: "And what then would human creativity be? Creativity ceases to be about self-expression. Instead it involves understanding (through research, analysis, intuition and so on) and then facilitating these various larger processes of creative emergence that constitute the many levels of evolution. Besides transcending self-expression, creativity then escapes the current frivolous obsessions with form and theory—a symptom of how lost we are and lacking in vision as to the purposes of architecture—to be about expanding the world of human possibility so that we can become more of who we aspire to be in our emerging view of what it is to be fully human."[14]

According to Buchanan, architectural education should include a series of balanced objective and subjective lecture courses regarding evolution, ecology, human ecology, and history of human settlements, to provide the necessary context for all forms of environmental design. Courses on climate and cultural adaptations to climate; courses on resource flow (materials, food, energy); and courses in psychology, perception, phenomenology, environmental psychology are essential to an inclusive education.[15]

I completely agree. Many practicing architects have benefited from a multidisciplinary education. More and more architectural schools are part of larger colleges which include visual arts, performing arts, industrial design, planning and landscape architecture. There are at least two schools which are engaging neuroscience in their programs—The New School of Architecture + Design (NSAD) in San Diego, and The University of Arizona (UofA) College of Architecture, Planning and Landscape Architecture (CAPLA). The NSAD is collaborating with ANFA, the University of California San Diego, and other organizations to offer courses. At the UofA, CAPLA and the Institute for Place and Well-Being are working together on research and developing programs and courses.

EMPATHY

The neuroscientist Jonas Kaplan participated in a neuroscience and architectural workshop on correctional facility design that took place in 2006. He was part of a team of UCLA researchers who were the first to make a direct recording of mirror neurons in the human brain. He wrote: "Mirror neurons, many say, are what make us human. They are the cells in the brain that fire not only when we perform a particular action but also when we watch someone else perform the same action. Neuroscientists believe this 'mirroring' is the mechanism by which we can 'read' the minds of others and empathize with them. It's how we 'feel' someone's pain, how we discern a grimace from a grin, a smirk from a smile."[16]

As architects, we discuss the importance of empathic connections to people, to other's experiences, and to nature. We strive to capitalize on human connection and design to encourage interaction and collaboration, and—although rarely specifically stated—empathy is implied throughout our work. Understanding empathy and mirroring could help us strengthen these connections. The use of transparent materials is one design strategy often employed to enhance human interaction, for example. What if, as the correctional facility design workshop hypothesized, the presence of a material (i.e., glass) actually weakened or eliminated the chance for mirroring, and therefore weakened the opportunity for empathy and "productive" interaction? A specific hypothesis generated in this discussion was whether an increase in pro-social inmate-staff contact would reduce inmate disorder because of an increased activity of mirror neurons. This behavior could be measured and different architectural interventions would be used, such as a transparent barrier and the elimination of a barrier.

DEFINING PERFORMANCE OUTCOMES

Knowing scientific outcomes could provide evidence for specific design goals. In May 2006, I attended a neuroscience laboratory design workshop held at the Dana Center in Washington, D.C. ANFA and the Dana Alliance for Brain Initiatives partnered to critically examine neuroscience laboratories and offices—to propose hypotheses about the cognitive processes affected by characteristics common to these laboratories and offices. Four basic workgroups generated hypotheses. The workgroup topics were as follows: (1) creativity/discovery; (2) productivity; (3) learning and memory; and (4) stress. The workshop was conducted so the groups would brainstorm about their topic and then come back to the big group with hypotheses, study designs, neuroscience and behavioral performance measurement techniques, and outcomes to be measured. Discussions around the creativity/discovery topic included inquiries such as: "How do you seek inspiration?" (online, hard-copy searches, communication, libraries, workshops); "What triggers your memory?"—is it seeing books in your library? Photos? Does space itself serve as an uninterrupted moment for memory? Does broadband noise influence learning and memory?

Do areas provided for presentation of artifacts gathered from previous experiences create cues to memory? Does daylight and/or daylight with views result in reduced cognitive fatigue and enhanced attention/clarity, permitting longer experiment cycles? Or does individual control of sensory aspects of ambient environment result in reduced stress, enhanced attention and productivity? One example hypothesis was: "Spaces planned for high interaction between lab members can prime individuals for problem-solving activities by controlling stress and stimulation levels." The design variables could include open atriums for high interaction and closed rooms for low. The behavioral performance would assess the frequency of interaction and the quality and quantity of ideas. Measurements would include traffic patterns and wayfinding for design; EEG and MEG levels after interaction period and the testing of right-side versus left-side brain activity and heart rate for neuroscience; and videography/motion studies and personal essays from the participants to measure the behaviors.

These hypotheses and questions demonstrate the type of inquiry required to begin understanding an architect's and neuroscientist's point of view; what makes a scientist more creative, inspired, and productive? Jonah Lehrer's *Proust Was a Neuroscientist* cites many examples of artistic intuition anticipating neuroscience. Walt Whitman discusses the proposition "the body is the soul," and understood the phantom limb phenomenon of soldiers who had had limbs amputated but still felt them. Auguste Escoffier, the chef who invented veal stock, understood the "secret of deliciousness"—umami. This taste was later determined to be the fifth taste (in addition to sweet, sour, bitter, and salty).[17] Architecture and neuroscience need to work together to generate scientific hypotheses based on experience and observations (from the artist/architect) that can then be studied using science.

APPLYING THE RESEARCH

In the following paragraphs, I will attempt to illustrate strategies of applying neuroscience research to design at several scales: individual (home); local (school); national (justice facilities and homeland security); and spiritual (cathedral). Since the neuroscience-architecture research is still in its infancy, some of the examples will be based on hypotheses pending study. To reinforce the proposal, recall the last sentence of Fred Gage's quote earlier in this chapter: "In planning the environments in which we live, architectural design changes our brain and our behavior."[18]

INDIVIDUAL SCALE: HOME

Home has different meanings for different people. For me, it is a place to rejuvenate, to be with family and friends, to feel safe, to prepare for the day, to celebrate, to find peace, to be creative, to remember. We purchased our home in 1994 and began renovations immediately. Our home, the Yia Yia Pavilion & Studio, is about investing in a

9.1 Farling studio, Phoenix, Arizona: interior.

community and making it stronger by reworking a 1950s ranch house into a long-term family home. My husband and I designed the home as a pavilion surrounded by gardens; each space has a distinct connection to the exterior, while the living-dining-cooking spaces can be transformed into a large outdoor "Arizona" room.

Our first major building addition was a master bedroom and studio. The studio is a space of contemplation, which opens to the north and captures a chunk of blue light and sky. Intuitively, we knew we wanted north light and views of nature, but I was also familiar with research which showed that reduced cognitive fatigue and enhanced attention and clarity would result from these intuitive design decisions. The studio sits above the roof level of the original residence and on top of the new master bedroom, bath, and laundry. The upper level is clad in plate aluminum, which reflects the sky—day or night. The lower level is clad in salvaged translucent plastic panels which allow for great light transmission while providing a high level of thermal insulation. The bedroom opens into and shares the trellis-covered west garden patio. In 2009 the original 1956 ranch house was transformed into the "Yia Yia Pavilion." Affectionately named after my Yia Yia (Greek for grandmother), the sunken indoor/outdoor living and dining room minimizes

9.2 Farling studio, Phoenix, Arizona: exterior.

the day-to-day footprint of air-conditioned space, and sliding the glass wall away turns it into a large outdoor room for those special occasions and holidays that Yia Yia loved. Initial and long-term consumption is controlled using solar energy, high-efficiency HVAC and appliances, ceiling fans, dimmable lighting, a biofuel fireplace niche, carefully controlled daylighting, very low-maintenance exterior materials, reused lumber, reused broken-up concrete pavers and ground cover, salvaged plastic wall panels, insulated glazing, natural ventilation, and low-water-use landscaping. The design of the house is about finding the balance between building performance and experiential pleasure. Reuse of materials was not only environmentally responsible—the materials hold memories for us. We lived in the house for almost sixteen years before our latest renovation—there are stories in the reused wood and concrete that trigger our memories and provide comfort and familiarity. New details such as our daughter's handprint in the concrete, or the shelves that hold personal mementos and books, or smells from the orange blossoms, also trigger memories that influence our creativity or make us feel comfortable. The design of a home should engage all the senses; neuroscience can potentially help us, as architects, understand these potentials more fully.

9.3 Farling residence, Phoenix, Arizona: Yia Yia Pavilion.

LOCAL SCALE: SCHOOL

A 2012 article in *Architectural Record* noted that the British government is calling for "simpler, rectilinear forms" in state schools with no indents, curves, glazed walls, roof terraces, etc. At least one Member of Parliament wanted the schools to be architect-free and was intent on saving money: yet another example of a lack of trust in architects.[19] The school environment profoundly affects the well-being, health, and learning of students, not to mention the impacts on faculty and staff. I heard a similar attitude last year at an orientation of a local middle-school preparatory academy. The headmaster said that as we could see, money is not spent on the facilities—money is spent on the programs. To most people this would sound reasonable. The classroom and auxiliary spaces he was referring to were windowless, beige-colored boxes with noisy mechanical equipment. There is already quite a bit of research to prove that natural daylight and views of nature improve math and reading scores for students,[20] but if we could definitively say why the windows and views are beneficial, perhaps the quality of school facilities would become a priority. At the 2012 "Minding Design" symposium at Taliesin West, Michael Arbib mentioned John P. Eberhard's statement "neuroscience will change our understanding of classroom design." This refers to questioning our standardized classrooms, which do not respond to developmental stages. Also, many classroom designs, especially in older facilities, do not account for negative impact on learning due to disturbing noise or discomfort (too hot or too cold). Dr. Gage hypothesized a potential positive design impact in his keynote theme address: "What is it about the stimulation that is occurring externally that is enhancing the students' ability to acquire new information? ... One can imagine that external stimulation, even in a classroom where students are concentrating and learning, could act as a general activator in certain brain areas, which in turn makes the brain more receptive to information coming in from the teacher."[21] Scientific evidence to support these claims could make it more difficult to deny the importance of seeking an architect who considers the individual student's, staff's, and faculty member's experience to be a serious priority. The headmaster might not brag about neglecting spatial qualities of the school. This neuroscience-architecture research still needs to be conducted, but would provide important information for the designer.

Another example of applying the research is currently being used in a local school district. I had the opportunity to work with Marlene Imirizian, FAIA, on the *Arizona School Design Primer: The Basic Elements of School Design*.[22] The primer was developed as a first step toward providing basic school design considerations, due to the lack of awareness in Arizona regarding the negative impacts of school buildings with little daylight, poor finishes, etc. As a follow-up to the primer, we are continuing the discussion with an elementary school district. This facility's staff firmly believe that their work can impact on learning and teaching in a positive manner—an unusual philosophy for a facilities department. The district has no capital plan for new school construction, but does have a large inventory of twenty-one existing schools of various ages. We have created a multidisciplinary design team to execute a toolkit with the goal of improving conditions for faculty and students. The toolkit focuses on specific strategies based on research in several categories. The district specifically asked for outcome-based solutions—for maintenance as well as new construction. One comment that stood out for me concerned comfort. When staff use leaf-blowers outside a classroom window, the noise disrupts some students with special needs; often it takes up to twenty minutes for the student to recover and refocus on the lesson. While this is not a building design issue, similar problems can occur with noisy mechanical systems or inappropriate adjacencies of programs.

From an ecological viewpoint, a school provides an extended family of teachers, friends, and staff. The school is fully immersed in the community. Often a school functions as a community center, and/or the playgrounds are used as community parks after hours. Schools are a continuation of the home. Children spend more time in school than in any other building other than their home. It is our responsibility to investigate the potentially profound influence of the design of our schools.

NATIONAL SCALE: JUSTICE FACILITIES AND HOMELAND SECURITY

Another focus for research has been correctional environments. Correctional facilities are "total environments" where inmates may spend long periods and are completely dependent on the institution for all their needs. Several years ago, I was part of a team whose objective was to evaluate stress of correctional officers in a jail. For the intervention, we introduced a large nature mural on two prominent and strategic walls in the booking area. Data were collected on the correctional staff before and after the intervention. The neuroscientists were able to analyze the data, and determined significant positive psychophysiological effects on measures of stress via heart rate (inter-beat intervals) variability. The heart rate monitoring was supplemented by additional measures of stress: the "backward digit span" (which tests mental agility and fatigue) and subjective questionnaires in which respondents gave their subjective assessment of stress levels.[23] Results of the data analysis demonstrated that, after installation of the mural, intake officers' heart rate was significantly lower at the beginning of the shift; the rate of increase of heart rate was significantly less from the beginning to end of the shift; and there was a significant increase in "log power," indicating inhibition of heart rate. The rate of log power increase is statistically significant and is consistent with lower heart rate and less stress. These data are suggestive of a pattern that is consistent with reduced stress at the end of the day after the exposure to the mural intervention.[24]

As I said above, building performance often takes precedence over human experience. One example which used neuroscience and architecture research was the design of a land port of entry on the Arizona-Mexico border. Land ports of entry are complex and rich with geographical, political, emotional, and cultural boundaries. Crossing these boundaries is highly sensitive for both those crossing and for those policing the border. For this project, I, with the other members of the Jones Studio design team, developed goals with the stakeholders. We gave top priority to security. This goal was problematic because it demands a reactive solution that responds only to the possibility of extreme behavior. Layers of fences, sight lines, and bullet-resistant materials govern. The design team sought to provide proactive measures by fully understanding all the underlying issues of the

9.4 Jones Studio, Mariposa Land Port of Entry, Nogales, Arizona.

users, including those of the visitors. The team concluded that reducing stress for officers at the port would have widespread benefits. To accomplish this, the design team applied the jail booking area research mentioned earlier. Easing stress allows for more efficient operations. Therefore, we prioritized views of nature as a technique to reduce stress and improve the cognitive performance of the officers. The port officers could relate to this premise—easing operations allows for a safer environment. The port design was focused around a literal oasis in the middle of the port, in addition to areas of respite and views of nature throughout. The oasis also provided a safe zone for the officers—providing a programmatic function as well as the benefits of the views of nature. We still provided layers of defense, but it was done as unobtrusively as possible—using buildings as perimeter in lieu of fences, etc. Without such evidence and logic, the design might not have been justified in the opinion of the officers. The reduced stress might also have a ripple effect on those visitors entering the port—for many, their first time in the United States—contributing to an enhanced experience for all users.

SPIRITUAL SCALE: CATHEDRAL

I would like to refer back to the beginning of this chapter: Eberhard's example of experiencing Amiens Cathedral. His description of the brain's activities while experiencing architecture tells the story of how we are moved. Most architects I know aspire to heal or increase quality of life, and raise spirits. Every moment we experience is precious, and we cannot afford to underestimate the power of our buildings and spaces impacting on people. Through the help of neuroscience, we can only help to increase society's expectations of the built environment, and elevate both the status and responsibility of the architectural profession.

NOTES

1. John P. Eberhard, "You Need to Know What You Don't Know," *AIArchitect* (January 2006):1.

2. For further information about the Academy of Neuroscience for Architecture, visit their website at: <http://www.anfarch.org>.

3. "Academy of Neuroscience for Architecture," Research Associates, <http://www.anfarch.org/people/associates>.

4. Fred Gage, "Neuroscience and Architecture," theme presentation, Convention Center, San Diego, May 9, 2003, 2–3. Emphasis added.

5. Peter Buchanan, "The Big Rethink: Towards a Complete Architecture," *Architectural Review* (December 21, 2011).

6. Kenneth Frampton, "Towards a Critical Regionalism: Six Points for an Architecture of Resistance," in Hal Foster, ed., *The Anti-Aesthetic: Essays on Postmodern Culture* (Seattle: Bay Press, 1983), 27.

7. Eve Edelstein and Peter Otto, "Reduction of Medical Error by Design: How the Neuroscience of Hearing Informs Healthcare Design," presentation, Salk Institute for Biological Sciences, La Jolla, CA, September 20, 2012.

8. Richard E. Wener, *The Environmental Psychology of Prisons and Jails: Creating Humane Spaces in Secure Settings* (New York: Cambridge University Press, 2012), 162.

9. Meredith Banasiak, "Gehry Talks about Architecture and the Mind at Neuroscience Conference," *AIA News Headlines*, <http://www.aia.org/aiarchitect/thisweek06/1110/1110n-gehry.cfm>.

10. Vitruvius, *The Ten Books on Architecture*, trans. Morris Hicky Morgan (New York: Dover, 1960).

11. Christopher Alexander et al., *A Pattern Language: Towns, Buildings, Construction* (New York: Oxford University Press, 1977), 58–62, 293–296, 377–380, 1047–1062, 1105–1107.

12. Frampton, "Towards a Critical Regionalism," 16.

13. Ibid., 28.

14. Peter Buchanan, "The Big Rethink: The Purposes of Architecture," *Architectural Review* (March 27, 2012).

15. Ibid.

16. Mark Wheeler, "UCLA Researchers Make First Direct Recording of Mirror Neurons in Human Brain," *UCLA Newsroom* (April 12, 2010).

17. Jonah Lehrer, *Proust Was a Neuroscientist* (New York: Houghton Mifflin, 2007), 1–24, 53–74.

18. Gage, "Neuroscience and Architecture," 3.

19. Christopher Turner, "Brits Declare War on School Curves," *Architectural Record*, vol. 200, issue 11 (October 17, 2012): 25.

20. Heschong Mahone Group, "Daylighting in Schools: An Investigation into the Relationships Between Daylighting and Human Performance," Pacific Gas and Electric Company, August 20, 1999; Heschong Mahone Group, "Windows and Classrooms: A Study of Student Performance and the Indoor Environment," California Energy Commission, October 2003.

21. Gage, "Neuroscience and Architecture," 3.

22. Marlene S. Imirizian, *Arizona School Design Primer: The Basic Elements of School Design* (Phoenix: Marlene S. Imirizian, 2013).

23. Jay Farbstein, Melissa Farling, and Richard E. Wener, *Effects of a Simulated Nature View on Cognitive and Psycho-physiological Responses of Correctional Officers in a Jail Intake Area* (Washington, D.C.: National Institute of Corrections, 2009).

24. Ibid., 19–21.

10

NEUROSCIENCE FOR ARCHITECTURE

Thomas D. Albright

Buildings serve many purposes. One might argue that their primary function is to provide shelter for the inhabitants and their possessions—a place to stay warm and dry, and to sleep without fear of predators or pathogens. Buildings also provide spaces to safely contain and facilitate social groups focused on learning, work, or play. And they provide for privacy, a space for solace and retreat from the social demands of human existence.

These primary physical requirements, and their many subsidiaries, simply reflect the fact that we are biological creatures. In addition to building constraints dictated by site, materials, and budget, an architect must respond to the nonnegotiable facts of human biology. Indeed, architecture has always bowed to biology: the countertop heights in kitchens, the rise:run ratio of stairs, lighting, water sources, heat and airflow through a building, are all patent solutions to salient biological needs and constraints. There are creative technology-based extensions of these solutions afoot in the form of smart homes. But there are subtler instances in which a deeper understanding of human biology affords a qualitatively superior solution. Consider, for example, the ascendance of the door lever as a design imperative imposed by biology. Seen from a strictly biomechanical perspective, a door lever is a far better tool than a traditional round doorknob for opening the latch. Pressure to adopt this superior solution came largely from recognition that it could benefit people with certain biological limitations ("physical disabilities"). Not surprisingly, the U.S. Americans with Disabilities Act (1990) has mandated the use of door levers because their design is easy to grasp with one hand and does not require "tight grasping or pinching or twisting of the wrist to operate."[1] Here is a case in which design centered explicitly on the details of a biological problem allows for greater accessibility and enhanced use.

At the same time that our buildings provide physical solutions to problems dictated by human biology, we also expect them to satisfy our psychological needs. We expect them to inspire and excite us, to promote mental states that lead us to discover, understand and create, to heal and find our way, to summon the better angels of our nature. We expect them to be beautiful. Not surprisingly, psychological considerations have been a part of the design process since humans began constructing lasting communal environments. The ancient tradition of Vaastu Veda, which dictated the design of temples and dwellings in early Hindu society, focused on ways in which a building directs "spiritual energies" that influence the souls of the inhabitants[2]—or, in today's parlance, the ways in which design influences the many facets of mental well-being. Feng shui, the ancient Chinese philosophy of building design, emerged for similar reasons.[3]

VAASTU VEDA IN THE AGE OF NEUROSCIENCE

While the basic psychological needs of a building's inhabitants today remain largely the same as they were in ancient times, we have one notable tool that promises a new perspective on how buildings influence our mental states: the modern field of neuroscience. Considered broadly, neuroscience is the umbrella for a collection of empirical disciplines—among them biology, experimental psychology, cognitive science, chemistry, anatomy, physiology, computer science—that investigate the relationship between the brain and behavior.[4] There are multiple internal processes that underlie that relationship, including sensation, perception, cognition, memory, and emotion.

There are also multiple levels at which we can investigate and characterize the relationship between brain and behavior. We can, for example, describe behavior in terms of the interactions between large brain systems for sensory processing and memory. Or we can drill down and explore how cellular interactions within circuits of brain cells (neurons) give rise to larger system properties, such as visual perception. Deeper still, we can explore the molecular components and events that underlie the behaviors of individual neurons, or the genetic codes and patterns of gene expression that produce the cellular substrates and organized circuits for brain function.

Most importantly, modern neuroscience affords the tools and concepts that enable us to identify the causal biological chains extending from genes to human behavior. This powerful approach, and the rich understanding of brain function that it affords, naturally has broad implications for and applications to many problems in human society, particularly in the field of medicine. But one might reasonably ask—and many do—whether there is any practical value for architecture and design that comes from knowing, for example, how neurons are wired up in the brain. I argue that there is value: knowing how the machine works can offer insights into its performance and limitations, insights into what it does best and how we might be able to tune it up for the task at hand. In the same way that understanding of an amplifier circuit in your car radio can lead to principled hypotheses regarding the types of sound it plays best, knowledge of how the human visual system is wired up may, for example, lead to unexpected predictions about the visual aesthetics or navigability of a building. At the same time, of course, the level of analysis of brain function should be appropriate for the question. In the same sense that knowledge of electron flow in a transistor offers few practical insights into what your radio is capable of, it seems unlikely that today's knowledge of patterns of gene expression that underlie brain circuits will yield much grist for the mill of design. That said, our understanding of brain development, function, and plasticity is still evolving, and we may find that the larger multilevel picture eventually leads to new ways of thinking.

THE BRAIN AS AN INFORMATION-PROCESSOR

In trying to understand more concretely how neuroscience might be relevant to design, it is useful to think of the brain as an information-processing device, which of course it is. Indeed, it is the most powerful information-processing device known to man. The brain acquires information about the world through the senses and then organizes, interprets, and integrates that information. The brain assigns value, affect, and potential utility to the acquired information, and stores that information by means of memory in order to access it at a later time. These memories of information received form the basis for future actions.

Thinking further along these lines, we can make the argument that architecture is a multifaceted source of information. The sensory appearance tells us how space is organized, and thus its utility and navigability. Similarly, the appearance and its relationship to intended function may be profoundly symbolic, conjuring up a broader view of the responsibility to the users of the space and their relationship to society. Prior experiences with the world will of course come into play in understanding the meaning of the space and how it might most effectively serve its intended purpose, or inspire other unintended uses. And, of course, information conveyed by our senses, considered in a symbolic and functional context, may be the source of strong aesthetic and emotional responses, including our perception of beauty.

Building on this information-processing perspective, we can begin to articulate a few basic principles about how knowledge of the brain may bear upon architectural design. These principles conveniently fall into categories of information *acquisition*, *organization*, and *use*. In terms of acquisition, the built environment should be optimized to neuronal constraints on sensory performance and information-seeking behavior, and optimized with respect to the adaptability of those constraints. At the simplest level, for example, knowing something about human visual sensitivity—what we see best and what we have difficulty seeing—may define rules for efficient design of environments for labor, learning, healing, and recreation. I will elaborate on some examples of optimizing sensory performance later in this chapter.

In terms of organization, the built environment should facilitate perceptual organization and engender the formation of cognitive schema/neuronal maps for the task at hand. An example of the relevance of neuronal maps can be found in research on wayfinding behavior.[5] A rich vein of neuroscience research has revealed much about how space, and the location of an observer in space, is represented by populations of neurons—neuronal maps of space—in a brain structure known as the hippocampus.[6] This knowledge, in conjunction with an understanding of how landmarks and other sensory cues in the built environment facilitate wayfinding, may lead to new ideas about how to facilitate navigability by design. These ideas, in turn, may help those who suffer from memory disorders associated with dementia, and help to improve design of transportation hubs and public areas in general.[7]

In terms of use, the built environment should elicit internal states that benefit sensory, perceptual, and cognitive performance and behavioral outcomes. "Internal states" here refers to those associated with focal attention, motivation, emotion, and stress. A number of recent studies support the plausible conjecture that certain environments elicit attentional states,[8] or states of anxiety and stress,[9] which can either facilitate or interfere with the ability of observers to respond to information embedded in the environment or to carry out actions for which the environment was intended. In work with Alzheimer's patients, for example, John Zeisel[10] has shown that architectural design elicits certain outcomes that have clinical value: anxiety and aggression are reduced in settings with greater privacy and personalization; social withdrawal is reduced in settings with limited numbers of common spaces that each have a distinctive identity; agitation is reduced in settings that are more residential than institutional in character. This type of knowledge could similarly inform the design of classrooms, lecture halls, health care facilities, workspaces, and more.

VISUAL FUNCTION, PERCEPTION, AND ARCHITECTURE

One area of neuroscience research that is particularly amenable to this kind of information-processing approach—and its relevance to architecture—is that associated with study of the visual system. This is true in part because vision plays a primary role in architectural experience, but also because we now have a wealth of information about how the visual system works.[11] In the following sections, I will highlight some examples drawn from our current understanding of vision, in order to illustrate the merits of this way of thinking. To set the stage, I will first briefly summarize the basic organization of the human visual system, as well as the neuroscience research methods used to study it.

Visual experience depends, of course, on information conveyed by patterns of light. Most of the patterned light that you see originates by reflectance from surfaces in your environment—sunlight returned from the façade of a building, for example. This reflected light is optically refracted by the crystalline lens in the front of your eye, yielding a focused image that is projected onto the back surface of the eye. This back surface is lined with a crucial neuronal tissue known as the retina, which is where phototransduction takes place: energy in the form of light is transduced into energy in the form of electrical signals, which are communicated by neurons. Retinal neurons carrying information in the form of such signals exit the eye via the optic nerve and terminate in a region near the center of the brain, known as the thalamus. Information reaching this stage is conveyed across chemical synapses and relayed on by thalamic fibers to reach the visual cortex. The visual cortex comprises the most posterior regions of the cerebral cortex, which is the large wrinkled sheet of neuronal tissue that forms the exterior surface of the human brain. The visual cortex is where high-level processing of visual images takes place, and it is the substrate that underlies our conscious visual experiences of the world. Our objective here is to understand how the organization of the visual cortex might have implications for the design of human environments.

EMPIRICAL APPROACHES TO UNDERSTANDING VISION

There is a variety of powerful experimental tools for studying the organization and function of the brain, which are summarized here as they apply to an understanding of the visual system.[12] Perhaps the simplest approach involves analysis of behavioral responses to sensory stimuli. This method, known as psychophysics, dates to the nineteenth century and involves asking people under very rigorous conditions to tell us what they observe when presented with visual stimuli that vary along simple dimensions, such as wavelength of light or direction of motion. From this we are able to precisely quantify what stimulus information observers are able to perceive, remember, and use to guide their actions. This approach is particularly valuable when used in conjunction with other experimental techniques, such as those that follow.

One important complement to psychophysics is neuroanatomy, which reveals the cellular units of brain function and their patterns of interconnections. With this approach we can, for example, trace the neuronal connections from the retina up through multiple stages of visual processing in the cerebral cortex, thereby yielding a wiring diagram of neuronal circuits.[13] Such wiring patterns reveal, in turn, computational principles by which visual information is combined and abstracted to yield perceptual experience.

Another powerful experimental technique is electrophysiology, the main goal of which is to understand how information flows through the system. To measure this flow, we use microelectrodes—fine wires that are insulated along their lengths and exposed at the very tips—that are inserted into the brain to monitor electrical signals (known as action potentials) from individual neurons. From such experiments we know that the frequency of electrical signals carried by a visual neuron is often correlated with a specific property of a visual stimulus. A neuron might thus "respond" selectively to a particular color of light, or to a specific shape.[14] These patterns of selective signaling reflect the visual information encoded by neuronal circuits. Moreover, by monitoring the ways in which signals are transformed from one processing stage to the next, we can infer the "goals" of each stage and gain insights into the underlying computation.

Fine-scale electrophysiology of the sort described above is largely restricted to use in experimental animals, but there are larger-scale approaches that involve assessment of patterns of brain activity recorded from the surface of the scalp. Despite the relative coarseness of the latter approach, electroencephalographic (EEG) methods are advantageous for our interest in architecture because they can be used to assess broad patterns of neuronal activity noninvasively in humans who are actively exploring an environment.[15]

Electrophysiological approaches are often complemented by a newer experimental technique known as functional magnetic resonance imaging (fMRI). This noninvasive method exploits the fact that: (1) oxygenated blood has a distinct signature in a magnetic resonance image, (2) oxygenated blood is dynamically redirected to regions of the brain that are metabolically active, and (3) neurons that are electrically active have a higher metabolic load. Thus the fMRI blood flow signal serves as a proxy for measurements of neuronal activity and can be used to identify brain regions that are active under different sensory, perceptual, cognitive and/or behavioral conditions.[16]

The various experimental techniques of modern neuroscience, summarized above, are most powerfully used in concert with one another, where they can collectively yield a rich and coherent picture of the ways in which information is acquired and organized by the brain, and used to make decisions and guide actions.

ON THE STATISTICAL PROPERTIES OF VISUAL INFORMATION

With this brief introduction to the organization of the visual system and the methods by which it can be studied, we can consider how current knowledge of information processing by the brain might suggest principles for design of human environments. I will begin with the premise that the brain has evolved to maximize acquisition of behaviorally relevant information about the environment, but must do so in the face of biological constraints. These constraints include various sources of noise and bottlenecks inherent to the neuronal machinery of the brain itself, the consequence of which is that our sensory systems are less than perfect transducers. Or, to put it more concretely, there are some things that we see better than others.

To illustrate how this limitation applies to architecture and design, we can start by measuring the physical properties of visual scenes from which the brain extracts information. There are many ways to do this—both natural and built environments have measurable statistics and we can quantify simple things like the frequency distributions of primary features, such as the different colors in a scene, or the orientations of contours (for example, those forming the frame of a window, or the branches of a tree). These simple statistics can be compared with the empirically determined sensitivity of the visual system for the same features, which provides a measure of the extent to which people can actually acquire (and thus use) certain classes of information present in the environment.

Employing the same approach, we can also quantify the statistics of higher-order image features—which are arguably more directly relevant to human behavior in natural and built environments—such as particular shapes and the joint probabilities of certain features (e.g., how often a specific color coincides in space with a certain shape). One specific example that has been looked at in some detail is the relationship between different line orientations as a function of their proximity in visual space.[17] As intuition suggests, there is a strong tendency for image contours that are nearby to have similar orientations, but as distance between them increases there is a progressive increase in the variance between pairs of contour orientations. One need only look at the contours of common man-made or natural objects—a teapot, for example, or a rose—to see that this distance-dependent contour orientation relationship simply reflects the physical properties of things in our visual world. The functional importance of this relationship can be seen by contrasting it with man-made objects that violate the principle: the image statistics of a Jackson Pollock painting,[18] for example, reflect a riot of angles and colors whose relationships yield no real perceptual synthesis.

FUNCTIONAL ORGANIZATION OF THE VISUAL BRAIN

Some unexpected insights and predictions come from consideration of image statistics in conjunction with knowledge of the organizational features of the visual cortex. Over the past few decades we have learned that there are a number different regions of the visual cortex that are specialized for the processing of unique types of visual information; one region processes contour orientation, another motion, another area processes color, and so on.[19] This knowledge has come, in part, from electrophysiological studies of the sort described above, in which the response (measured as frequency of action potentials) of a given visual neuron varies with the value of a simple stimulus along a specific feature dimension: for example, the particular angle of an oriented contour, or the particular direction of a moving pattern.

Figure 10.1 illustrates this type of cellular "tuning" as originally discovered for neurons in primary visual cortex.[20] The data represent action potentials recorded as a function of the orientation and direction of motion of a simple visual stimulus (an oriented contour). In this case, the recorded neuron responded best to a slightly off-vertical orientation moving up to the right, and the neuronal response waned as a function of the angular deviation of the contour relative to this preferred orientation. The vast majority of neurons in the primary visual cortex exhibit this property of "orientation selectivity." Their discovery in the 1960s by David Hubel and Torsten Wiesel transformed the way we understand the visual system, and fostered the development of a whole new set of techniques to study it. The existence of this specialized population of neurons in the cerebral cortex, and other populations that represent stimulus direction[21] and color,[22] accounts for the primacy of such simple features in our visual experience of the world.

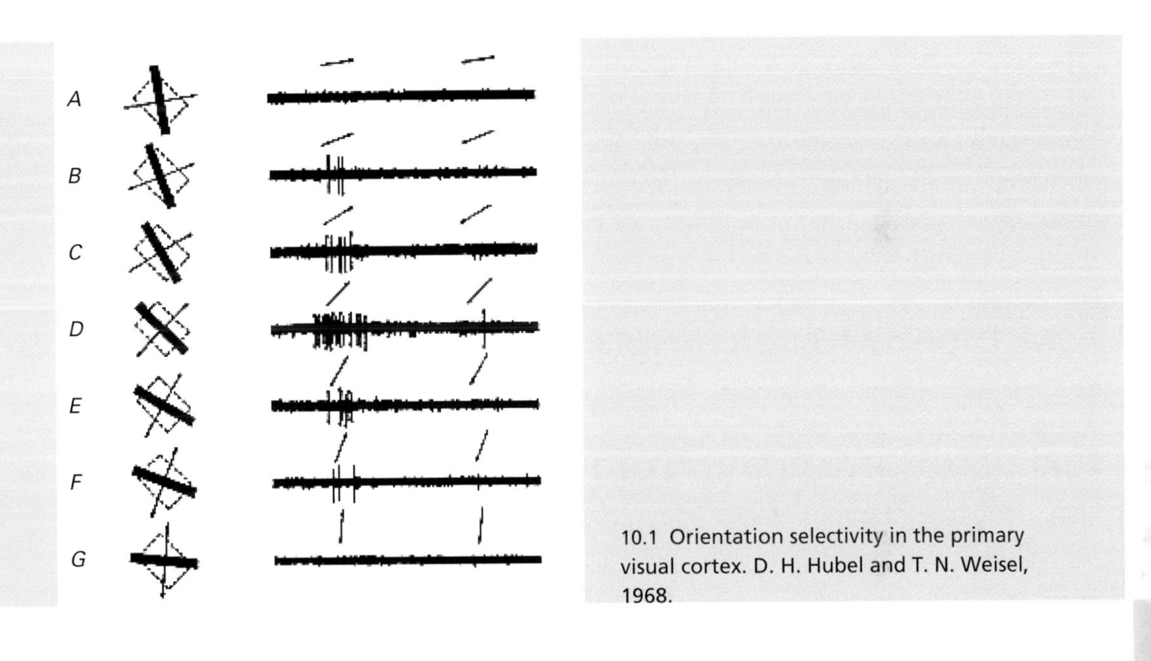

10.1 Orientation selectivity in the primary visual cortex. D. H. Hubel and T. N. Weisel, 1968.

Each of these functionally specific areas is further arranged according to certain organizational principles. One of these is columnar organization, which means that similar values of a given feature dimension (such as contour orientation or direction of motion) are represented in adjacent cortical tissues.[23] These functional columns extend through the thickness of the cerebral cortex and are mediated by neuronal microcircuits that correspond anatomically to the functional columns.[24] The neuronal architecture is such that the preferred value of the relevant feature (e.g., the preferred contour orientation) remains constant as one moves from the surface through the depth of the cortex, but changes gradually as one moves in the orthogonal plane, i.e., parallel to the cortical surface.[25] The scale of this system is fine, with a complete cycle of preferred orientations contained within less than a millimeter of cortex. A highly similar columnar system exists in a region of visual cortex specialized for encoding direction of motion.[26] In this case, the individual neurons represent specific directions, rather than contour orientations, and a complete cycle of direction columns similarly spans a region of cortex less than a millimeter across.

Another organizational principle of the visual system is built around the concept of association fields.[27] Association fields reflect patterns of local anatomical connections that link neurons representing specific values of a visual feature dimension. In the primary visual cortex, the specificity of these connections is made possible by the existence of an organized columnar system for representing contour orientations (see above). The connections are manifested as anatomical links between columns representing specific contour orientations. In particular, within cortical regions representing close-by locations in visual space, there exist strong connections between columns that represent similar orientations and only weak connections between columns that represent widely different orientations (perpendicular being the extreme).[28] As the spatial distance grows, the pattern of anatomical connections becomes more isotropic.

ASPECTS OF PERCEPTION FACILITATED BY NEURONAL ARCHITECTURE

These highly specific organizational properties for representing information about the visual environment raise interesting questions and conjectures about their relationship to visual perception. For one, we note that there is an apparent symmetry between the association fields for contour orientation and the statistics (summarized above) of contour orientations in the visual world. As we have seen, contours that are nearby in visual space are more commonly similar in orientation, relative to those that are distant in visual space. Analogously, in the visual cortex, cells representing similar orientations are preferentially interconnected provided that they also represent nearby locations in visual space. There are evolutionary arguments one can make: it seems highly likely that this cortical system for organizing visual information conferred a selective advantage for detecting statistical regularities in the world in which we evolved. At any rate, we hypothesize that the existence of the system helps to facilitate the processing of commonly occurring relationships between visual features.

A key part of this conjecture, which has implications for architecture and design, is the word *facilitate*. Human psychophysical experiments have shown, for example, that when people view random patterns of line segments, any colinear, or nearly colinear, relationships within those patterns tend to stand out perceptually from a background of

10.2 Field of wheat.

10.3 Green bodhi leaves.

10.4 Alaskan tundra.

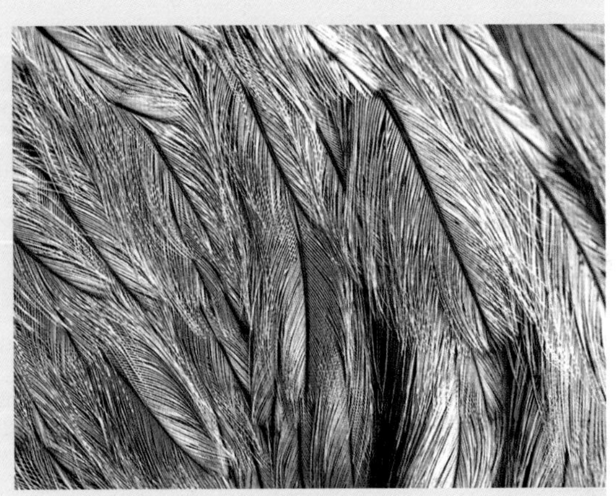

10.5 Feathers of an ostrich.

noise[29]—according to our hypothesis, perceptual sensitivity to these arrangements is *facilitated* by the organizational properties of the visual cortex.

As implied by the foregoing arguments, visual patterns in which there is a statistical regularity between adjacent contour orientations—repeating lines in colinear, curvilinear, parallel and radial patterns, for example—are ubiquitous in the natural world. Fields of grass, waves in the ocean, the veins of a leaf, the branches of a tree, the leaflets of a palm frond, or the barbs of a feather are all commonly encountered examples that embody this principle.

We hypothesize that man-made designs that adopt this same principle "benefit" in some way—detection of them is "facilitated"—by tapping into the highly organized neuronal

10.6 Fay Jones, Thorncrown Chapel, Fayetteville, Arkansas.

system for representing contour orientations. One need not look hard to find prized exemplars in the built environment that feature colinear, curvilinear, parallel and radial patterns: Fay Jones's Thorncrown Chapel in Fayetteville, Arkansas, the colonnades in Romanesque churches and monasteries such as the abbey at Assisi, or the rose window in the cathedral of Notre Dame. The cable-stayed bridge, which is commonly constructed using radial fans of cables to cantilever the road bed, is a particularly notable example. This is the most commonly built highway bridge today. There are many reasons for this that stem from advances in materials science and engineering, as well as economy of construction. But I speculate that the popularity of the cable-stayed bridge is also due, in part, to the fact that the gradually changing contours tap into something fundamental in the native organization of our visual system. There is, I will argue, an attractiveness to

10.7 Cloisters, Monreale, Sicily.

these designs that originates from the ease with which they are processed and perceived by our visual systems.

THE SENSE OF ORDER

Neuroscientists were not the first to make this connection. Ernst Gombrich, one of the great geniuses of twentieth-century arts and humanities, wrote and reflected deeply on the relationship between art and visual perception.[30] His text *The Sense of Order: A Study in the Psychology of Decorative Art* addresses the use of certain timeless design features in art and architecture. Summarizing his thesis elsewhere, Gombrich wrote: "I claim that the formal characteristics of most human products, from tools to buildings and from clothing to ornament, can be seen as manifestations of that sense of order which is deeply rooted in man's biological heritage. These ordered events in our environment which exhibit rhythmical or other regular features (the waves of the sea or the uniform texture of a cornfield) easily 'lock in' with our tentative projections of order and thereby sink below the threshold of our attention while any change in these regularities leads to an arousal of attention. Hence the artificial environment man has created for himself satisfies the dual demand for easy adjustment and easy arousal."[31]

Gombrich was not a neuroscientist, of course, but his concept of "manifestations of that sense of order which is deeply rooted in man's biological heritage" and his suggestion that "these ordered events in our environment ... easily 'lock in' with our tentative projections of order" resonate deeply with the view that our perception of the world depends heavily upon highly ordered neurobiological characteristics of the human visual system. Again without knowledge of the neuroscience of vision, Gombrich expanded along similar lines: "There is an observable bias in our perception for simple configurations, straight lines, circles and other simple orders and we will tend to see such regularities rather than random shapes and our encounter with the chaotic world outside. Just as scattered iron filings in a magnetic field order themselves into a pattern, so the nervous impulses reaching the visual cortex are subject to the forces of attraction and repulsion."[32] Gombrich's iron filings metaphor is striking in the present context, as it poetically captures the notion that the organizational properties of the visual system serve to efficiently encode statistical regularities in the visual world.

Gombrich spoke at length about designs that impart this sense of order. Some examples include the mosaics at the Alhambra, and the paper and textile patterns of William Morris. To these I would add the decorative designs of Frank Lloyd Wright from a similar period to those of Morris. For each of these examples, it is not necessary to sit and examine how it is put together; you see one part and a perceptual understanding of the whole follows without visual scrutiny—they are repetitive designs that capitalize on the ordered nature of the visual cortex.

Similar arguments apply to mandalas, which have been used as meditation aids for centuries in the spiritual practices of Hinduism and Tibetan Buddhism. As for the decorative patterns cited above, mandalas have image statistics that are complementary to the organization of the visual cortex. Our conjecture is that they have an ordering effect owing to the ease of visual processing—they are calming, regular structures.

10.8 Frank Lloyd Wright, textile block pattern.

10.9 Frank Lloyd Wright, textile block house, Los Angeles.

10.10 Rose window at Notre Dame, Paris.

By the same logic, of course, we should expect that images possessing irregular statistics, or properties that do not tap into to the organizational features of the visual cortex, should require greater effort to process and may lead to confusion, disturbance, and distraction.

FAMILIARITY VERSUS NOVELTY

I interpret Gombrich's statement that the built environment "satisfies the dual demand for easy adjustment and easy arousal" to mean that the optimal environment has varying degrees of familiarity and novelty. That is, we create features in our environment with a sense of order; of things that are familiar. Without visual scrutiny, such features are

10.11 Tibetan sand mandala. Minneapolis Institute of Arts.

easily processed because they tap into the inherent organization of our brain systems for visual perception. This order provides a suitable background—and liberates neuronal resources—for detection of novelty (a predator or an intruder, perhaps, or a new piece of furniture), which is nearly always of behavioral significance and demanding of attention. To put it simply, the built environment tends to reflect the way visual perception works.

Gombrich was not the only person who noticed this phenomenon. Oscar Wilde also observed: "The art that is frankly decorative is the art to live with. The harmony that resides in the delicate proportions of lines and masses becomes mirrored in the mind. The repetitions of pattern give us rest. Decorative art prepares the soul for the reception of imaginative work."[33] Again, Wilde is using literary language to describe how the visual system functions between the poles of familiarity and novelty. Repetition gives us rest, because we are not required to scrutinize every part of it. Comfort derives from ease of visual processing. Wilde suggests that the regularity of background sets the stage for truly imaginative work, for something new to emerge.

BRAIN AND BEAUTY

It should not go unnoticed that these ideas have implications for the neurobiology and evolution of aesthetics. There are surely many different reasons for the aesthetic judgments that we make about features of the natural and built environment, many reasons why we find beauty in one form and ugliness in another. Much of this is cultural and learned. Doubtless many people will tell you that Leonardo's *Mona Lisa* is beautiful, simply because that is what we have taught them. Oftentimes judgments of beauty will reflect frequent exposure to certain stimuli in the presence of reward (money, information, social power, or sex), or a cultural "consensus" defined by commercial interests and displayed through magazines, billboards, and television. But the foregoing discussion suggests a definition of beauty based on ease of visual processing—beauty defined by the extent to which features of the visual environment engage organized processing structures in the visual brain, and are thus readily acquired, organized, and "understood." Evolution is invoked in this definition of beauty, since we hypothesize that the relevant brain structures exist because they conferred a selective advantage for survival and reproduction in an environment replete with the image statistics described herein.

GENERALITY OF PRINCIPLES

The latter part of this chapter has focused on a specific set of organizational features in the visual cortex—those involved in the detection and representation of oriented contours—primarily because this is the visual submodality that we know the most about. The principles exemplified by this submodality are likely to be very general, however. Indeed, there are good reasons to believe that a detailed understanding of the architecture and function of brain systems for other visual submodalities (e.g., color or visual motion processing), or for other sensory modalities (e.g., audition and touch), will have similar implications for understanding the built environment.

PLASTICITY AND VISUAL ATTUNEMENT

Finally, it is important to note that the information-processing features of our brains are not rigid over time. On the contrary, they are plastic and tunable by experience. Recent evidence indicates that the sensitivities of our sensory systems are adapted to the statistics of our environment, but those sensitivities may change—they may be recalibrated—when the properties of the world change.[34] This adaptability has profound implications for design. Suppose, for example, that I adapt you to the baroque opulence of Marie Antoinette's bedroom in Versailles, and then move you to a minimalist home designed by Mies van der Rohe. The transition will elicit recalibration and will, we hypothesize, necessarily involve windows of time in which sensitivity is nonoptimal for the new environment. These considerations have particularly important implications for the design of spaces for work and learning, as frequent changes of environmental statistics may interfere with the ability of observers to acquire, organize, and use information from the environment.

CONCLUSIONS: TOWARD A NEUROSCIENCE FOR ARCHITECTURE

Neuroscience is a new research discipline in the armament of longstanding efforts to understand the influence of built environments over human mental function and behavior. Using a variety of powerful experimental approaches, and focusing efforts on the information-processing capacities of the brain, we have begun to develop an empirical understanding of how design features influence the acquisition, organization, and use of information present in the built environment. On the basis of this understanding, we argue that selective pressures over the course of human evolution have yielded a visual brain that has highly specific and tunable organizational properties for representing key statistics of the environment, such as commonly occurring features and conjunctions of features. Simple visual pattern types, which are commonly used in architectural and decorative design, mirror these environmental statistics. These patterns are readily "seen" without scrutiny, yielding a "sense of order" because they tap into existing neuronal substrates. A fuller understanding of these relationships between organizational properties of the brain and visual environmental statistics may lead to novel design principles.

NOTES

1. Americans with Disabilities Act, 1990.

2. S. Anath, *The Penguin Guide to Vaastu: The Classical Indian Science of Architecture and Design* (New Delhi: Penguin Books India, 1999).

3. S. Skinner, *Feng Shui History: The Story of Classical Feng Shui in China and the West from 221 BC to 2012 AD* (Singapore: Golden Hoard Press, 2012).

4. E. Kandel, J. H. Schwartz, T. M. Jessell, S. A. Siegelbaum, and A. J. Hudspeth, eds., *Principles of Neural Science*, 5th edn. (New York: McGraw-Hill, 2012).

5. R. G. Golledge, "Human Cognitive Maps and Wayfinding," in R. G. Golledge, ed., *Wayfinding Behavior* (Baltimore: Johns Hopkins University Press, 1999).

6. T. Hartley, C. Lever, N. Burgess, and J. O'Keefe, "Space in the Brain: How the Hippocampal Formation Supports Spatial Cognition," *Philosophical Transactions of the Royal Society, B* 369 (2014): 20120510.

7. Eduardo Macagno, "Research Technology and Architectural Design," lecture at ANFA conference, September 20, 2012, Salk Institute for Biological Studies, La Jolla, CA.

8. S. Kaplan, "The Restorative Benefits of Nature: Toward an Integrative Framework," *Journal of Environmental Psychology* 15 (1995): 159–182.

9. E. M. Sternberg, *Healing Spaces: The Science of Place and Well-Being* (Cambridge, MA: Belknap Press of Harvard University Press, 2010).

10. J. Zeisel, *I'm Still Here: A Breakthrough Approach to Living with Someone with Alzheimer's* (New York: Avery, 2009).

11. C. D. Gilbert, "The Constructive Nature of Visual Processing"; M. Meister and M. Tessier-Lavigne, "Low-Level Visual Processing: The Retina"; C. D. Gilbert, "Intermediate-Level Visual Processing and Visual Primitives"; T. D. Albright, "High-Level Visual Processing: Cognitive Influences"; and M. E. Goldberg and R. H. Wurtz, "Visual Processing and Action"; all in Kandel et al., *Principles of Neural Science*.

12. T. D. Albright, T. M. Jessell, E. Kandel, and M. I. Posner, "Neural Science: A Century of Progress and the Mysteries that Remain," *Cell* 100/*Neuron* 25 (2000): Supplement S1–S55.

13. D. J. Felleman and D. C. Van Essen, "Distributed Hierarchical Processing in the Primate Cerebral Cortex," *Cerebral Cortex* 1 (1991): 1–47.

14. Gilbert, "Intermediate-Level Visual Processing and Visual Primitives"; Albright, "High-Level Visual Processing: Cognitive Influences."

15. L. Zhang, Y. Chi, E. Edelstein, J. Schulze, K. Gramann, A. Velasquez, G. Cauwenberghs, and E. Macagno, "Wireless Physiological Monitoring and Ocular Tracking: 3D Calibration in a Fully-Immersive Virtual Health Care Environment," Proceedings of the IEEE Engineering in Medicine and Biology Conference, Buenos Aires, August 31–September 4, 2010, 4464–4467.

16. N. K. Logothetis, "What We Can Do and What We Cannot Do with fMRI," *Nature* 453 (2008): 869–878.

17. G. A. Cecchi, A. R. Rao, Y. Xiao, and E. Kaplan, "Statistics of Natural Scenes and Cortical Color Processing," *Journal of Vision* 10 (2010):1–13; W. S. Geisler, "Visual Perception and the Statistical Properties of Natural Scenes," *Annual Review of Neuroscience* 55 (2008): 167–192; M. I. Sigman, G. A. Cecchi, C. D. Gilbert, and M. O. Magnasco, "On a Common Circle: Natural Scenes and Gestalt Rules," *Proceedings of the National Academy of Sciences* 98 (2001):1935–1940.

18. Cecchi, Rao, Xiao, and Kaplan, "Statistics of Natural Scenes and Cortical Color Processing."

19. S. Zeki, "Parallelism and Functional Specialization in Human Visual Cortex," *Cold Spring Harbor Symposium on Quantitative Biology* 55 (1990): 651–661.

20. D. H. Hubel and T. N. Wiesel, "Receptive Fields and Functional Architecture of Monkey Striate Cortex," *Journal of Physiology* 195 (1968): 215–243.

21. T. D. Albright, "Cortical Processing of Visual Motion," in F. A. Miles and J. Wallman, eds., *Visual Motion and Its Role in the Stabilization of Gaze* (Amsterdam: Elsevier, 1993).

22. S. Zeki and L. Marini, "Three Stages of Colour Processing in the Human Brain," *Brain* 121 (1998): 1669–1685; K .R. Gegenfurtner, "Cortical Mechanisms of Colour Vision," *Nature Reviews Neuroscience* 4 (2003): 563–572.

23. V. B. Mountcastle, "An Organizing Principle for Cerebral Function: The Unit Model and the Distributed System," in G. M. Edelman and V. B. Mountcastle, eds., *The Mindful Brain* (Cambridge, MA: MIT Press, 1978); D. H. Hubel and T. N. Wiesel, "Sequence Regularity and Geometry of Orientation Columns in the Monkey Striate Cortex," *Journal of Comparative Neurology* 158 (1974): 267–293.

24. Mountcastle, "An Organizing Principle for Cerebral Function"; R. J. Douglas and K. A. C. Martin, "Neuronal Circuits of the Neocortex," *Annual Review of Neuroscience* 27 (2004): 419–451.

25. Hubel and Wiesel, "Sequence Regularity and Geometry of Orientation Columns in the Monkey Striate Cortex."

26. T. D. Albright, R. Desimone, and G. G. Gross, "Columnar Organization of Directionally Selective Cells in Visual Area MT of the Macaque," *Journal of Neurophysiology* 51 (1984): 16–31.

27. D. J. Field, A. Hayes, and R. Hess, "Contour Integration by the Human Visual System: Evidence for a Local 'Association Field,'" *Vision Research* 33 (1993): 173–193.

28. D. D. Stettler, A. Das, J. Bennett, and C. D. Gilbert, "Lateral Connectivity and Contextual Interactions in Macaque Primary Visual Cortex," *Neuron* 36 (2002): 739–750.

29. W. Li and C. D. Gilbert, "Global Contour Saliency and Local Colinear Interactions," *Journal of Neurophysiology* 88 (2002): 2846–2856; U. Polat and D. Sagi, "The Architecture of Perceptual Spatial Interactions," *Vision Research* 34 (1994): 73–78.

30. E. H. Gombrich, *Art and Illusion: A Study in the Psychology of Pictorial Representation* (Princeton: Princeton University Press, 1961).

31. E. H. Gombrich, "The Sense of Order: An Exchange," *New York Review of Books*, September 27, 1979.

32. E. H. Gombrich, *The Sense of Order: A Study in the Psychology of Decorative Art* (Ithaca: Cornell University Press, 1984), 4.

33. Oscar Wilde, "The Critic as Artist," 1913.

34. S. Gepshtein, L. A. Lesmes, and T. D. Albright, "Sensory Adaptation as Optimal Resource Allocation," *Proceedings of the National Academy of Sciences* 110 (2013): 4368–4373.

11

MOOD AND MEANING IN ARCHITECTURE

Alberto Pérez-Gómez

MIND: FROM ROMANTICISM TO NEUROPHENOMENOLOGY

In my book *Architecture and the Crisis of Modern Science*, I described how Western architecture was profoundly affected by the scientific revolution of the seventeenth century, revealing a set of intentions that were wholly modern long before the material changes brought about by the Industrial Revolution.[1] In relation to perception and cognition, an initial consequence of that momentous transformation in European thinking was the incorporation of René Descartes's dualistic epistemology/psychology into an understanding of how architecture communicates. This assumption had far-reaching consequences, opening the door for a subsequent understanding of architecture as a "sign," whose meaning was articulated as the intellectual "judgment" of exclusively visual qualities: the primary assumption of many twentieth-century poststructuralist and deconstructivist philosophers and architects, and one still present, often tacitly, among contemporary theoreticians.

The Cartesian understanding of cognition first appeared in architectural theory toward the end of the seventeenth century in the writings of Claude Perrault, the famous architect, medical doctor, biologist, and theoretician. He believed that architecture communicates its meanings to a disembodied soul (today often still identified with a brain, understood as the exclusive seat of consciousness), thoroughly bypassing the body, with its complex feelings and emotions.[2] Perrault assumed perception to be passive and cognition to be merely the result of the association of concepts and images in the brain. Like Descartes, Perrault believed that human consciousness (enabled by the pineal gland at the back of the head, conceived as a geometric and monocular point of contact between the measurable, intelligible world—*res extensa*—and the disembodied, rational soul—*res cogitans*) was capable of a perspectival visual perception, one that assured the human capacity to grasp the immutable geometric and mathematical truth of the external world.[3] He could question, for the first time ever in the history of architectural theory, the bodily experience of "harmony" as synesthetic, applicable to both hearing and sight embedded in kinesthesia: a phenomenon that had always been taken for granted since classical antiquity, and constituted the primary quality to be observed in architectural design. For Perrault, sight and hearing were autonomous and segregated receptors; therefore the inveterate experience of harmony in architecture was a fallacy, or at best the result of misguided associations between self-evident visual qualities and cultural assumptions.

While mainstream, technologically driven planning and architectural practice has remained caught in this framework of understanding until our very own times, around 150 years after Descartes's influential writings another, often unacknowledged revolution in the human sciences took place. Even though it was originally qualified as a mere reaction to positive reason, associated with the arts as they lost their claim to truth, and sometimes taken as a plea for "irrationality," over the last two centuries this transformation has proven to be as important for Western thought as the Galilean revolution.[4] This momentous shift happened at the end of the eighteenth century with the rise of romantic philosophy. Writers associated with this position questioned the dualism of Cartesian philosophy, and argued for the reciprocity and coemergence of inner and outer realms of human experience.[5] This initial insight allowed thinkers to establish a distance to materialism, assuming a critical position with regard to the technological dogma of their own times, while affirming the importance of imagination and the truth value of fiction. In his *Essais* (1795), Friedrich Schelling declares that it is our prerogative to question the times we live in and contemplate within ourselves eternity, with its immutable form. This is the only way to access our most precious certainties, to know "that anything is in the true sense of being, while the rest is only appearance." This intuition appears to us whenever we stop being an object for ourselves—we are not "in" linear time. Rather, "time, or pure eternity, is in us." This insight anticipates Maurice Merleau-Ponty's phenomenological understanding of time as thick present,[6] an experience that—I will argue below—is now corroborated by recent neurobiology. It is important to emphasize that Schelling added an important observation that qualifies his introspective critical understanding: "Even the most abstract notions retrieve an experience of life and existence ... *all our knowledge has as a point of departure direct experience.*"[7]

Recovering an insight that had been put forward initially by Aristotle in *De anima*, these romantic philosophers posited a concept of self which first feels and then thinks; the *I* who wakes up every morning is not equivalent to the Cartesian ego (an *I* that can believe itself existing only because he/she thinks).[8] The first person in romantic philosophy is always the same throughout her life, yet never fully "coincidental" with her thoughts. Her words point toward meanings, but never exhaust them. This embodied, nondualistic understanding of reality includes our emotions and feelings; its primary seat of awareness is *Gemüt*, and its most significant experience is *Stimmung*: attunement, understood as a search for lost integrity, health, wholeness and holiness. This concept has been shown to have its roots in traditional ideas about harmony (proportion) and temperance in the context of ancient classical and Renaissance cosmology, philosophy, music, and architectural theories,[9] eventually becoming cast as "atmosphere" or "mood"; a concept that is now understood to be of great consequence for art and architecture. The self is endowed with a consciousness that cannot be reduced to transparent reason, and since the elements of consciousness (subject, object, and action) are inevitably codependent, it starts to appear "ungrounded." Not surprisingly, romantic thinkers were fascinated by Eastern philosophy and started to incorporate some insights of Buddhism into their own positions, an approach welcomed eventually by Heidegger and more recently by enactive cognitive science.[10] They also could imagine a holistic biology that included the mind in the living body as opposed to the mechanistic medicine at the origins of contemporary physiology.

Romantic philosophy questioned positivistic thinking through narrative, giving rise to the modern novel as the privileged "place" both for the expression of *Stimmung* and for meditation on philosophical and ethical topics.[11] It also gave rise to the new discipline of history as interpretation (hermeneutics), distinct from the models and methodologies of the hard sciences, postulating this discipline as the proper mode of discourse for understanding human problems. This argument was expressed with clarity by Friedrich

Nietzsche in his crucial essay on "The Advantages and Disadvantages of History for Life," a text which is as relevant today as when it was first published.[12] I would argue that these positions were the precursors of late-nineteenth-century American pragmatism (William James and John Dewey), and of the early- and mid-twentieth-century phenomenology of Edmund Husserl and Maurice Merleau-Ponty, and thus lay at the root of later developments in American philosophy, like the contemporary work of Mark Johnson; of contemporary existential phenomenology; and also of the recent revolution in cognitive science that has built upon insights found in the above-mentioned philosophical positions, especially in the works of Evan Thompson and Alva Noë.

Given this lineage, I would like to suggest that from the point of view of Western architecture (whose assumptions, both instrumental and critical, are often universalized in our global village), the crucial moment when neuroscience starts to become useful for architects is after the now-famous "invention" of neurophenomenology in *The Embodied Mind* (1991), by Francisco Varela, Evan Thompson, and Eleanor Rosch. In a later work, Thompson explains how cognitive science came into being in the 1950s as a revolution against behaviorist psychology:[13] the same concern that motivated Maurice Merleau-Ponty to continue the work of his teacher Edmund Husserl in *Phenomenology of Perception* (first published in 1945). Early cognitivism, however, had as its central hypothesis the computer model of the mind. While cognitivism made meaning—in the sense of representational semantics—scientifically acceptable, it fundamentally banished consciousness from the science of the mind.[14] It soon became evident that abstract computation was not well suited to model the thought processes within the individual; this led in the 1980s to what has been labeled the "connectionist criticism," which focused on the neurological implausibility of the previous model.[15] While cognitivism still presumed the mind to be firmly bounded by the skull (cf. Descartes's psychology), connectionism started to offer a more dynamic understanding of the relationships between cognitive processes and the environment, creating models of such processes that took the form of artificial neural networks run as virtual systems on a digital computer.[16] These systems, however, did not involve any sensory and motor coupling with the environment; their inputs and outputs were artificial. Only "embodied dynamicism," the most recent approach of cognitive science that arose only in the 1990s, involved a true critical stance toward computationalism of any form.

Indeed, this latest approach in cognitive science stopped depending on analytic philosophy and computer brain models and started acknowledging the relations between cognitive processes and the real world. Embodied dynamicism called into question the conception of cognition as disembodied and abstract mental representation.[17] The mind and the world are simply *not* separate and independent of each other; the mind is an embodied dynamic system *in* the world, rather than merely a neural network in the head. For Varela, Thompson, and Rosch (1991), cognition is the exercise of skillful know-how in embodied and situated action, and cannot be reduced to prespecified problem-solving. In other words, the perceiver (subject), the perception, and the thing perceived (object) could never be said to exist independently; they are always codependent and coemergent, and therefore *ultimately* groundless or "empty" (a term taken by the authors from Mahayana Buddhism, to emphasize that this awareness is as opposed to absolutism as it is to a despairing nihilism, for out of the experience of emptiness in Buddhist meditation—letting go of grasping and anxiety—arises "sense" and mindful compassion). In that same seminal book they introduced the concept of cognition as "enaction," linking biological autopoiesis (living beings are autonomous agents that actively generate and maintain themselves) with the emergence of cognitive domains. The nervous system of all living beings in this view does not process information like a computer but rather creates meaning, *i.e., the perception of purpose in life*, whose articulation becomes more sophisticated with the acquisition of language in higher animals.

The world in this model is not a prespecified external realm represented externally by the brain, but a relational domain enacted by a being's particular mode of coupling with the environment. Experience in this approach is not a secondary issue (as it had been since Descartes), but becomes central to the understanding of the mind, and requires careful examination in the manner of phenomenology. In this connection, I would like to cite the work of distinguished neuroscientist Antonio Damasio, who has argued for the importance of emotions and feelings as essential building blocks of cognition, supporting human survival and enabling the spirit's greatest creations.[18] Recovering Baruch Spinoza's (and later phenomenology's) refusal to separate mind and body, Damasio has shown the continuity between emotions and appetites, feelings and concepts. He points out that every emotion is a variation of pleasure and pain, a condition of consciousness at the cellular level, always seeking homeostatic equilibrium.

In *Mind in Life*, Thompson relies upon the findings of Husserl and Merleau-Ponty to explicate selfhood and subjectivity from the ground up, accounting for the autonomy proper to living and cognitive beings. There is a deep convergence between phenomenology and the enactive approach that concerns the actual experience of time prefigured by romantic philosophy, and discussed by Merleau-Ponty in relation to his concept of *écart* as a "thick present." Thompson summarizes: "*The present moment manifests as a zone or span of actuality, instead of as an instantaneous flash*, thanks to the way our consciousness is structured. [It] manifests this way because of the nonlinear dynamics of brain activity. Weaving together these two types of analysis, the phenomenological and the neurobiological, in order to bridge the gap between subjective experience and biology, defines the aim of neurophenomenology."[19]

The consequences of this revolution in cognitive science are far-reaching, and the first two decades of the twenty-first century are witnessing the publication of important works exploring different aspects.[20] Alva Noë popularized the enactive understanding of perception and cognition in *Out of Our Heads: Why You Are Not Your Brain and Other Lessons from the Biology of Consciousness*, emphasizing particularly that in order to understand consciousness in humans and animals we must look not inward, but rather to the ways in which a whole animal goes on living in and responds to their world. Noë's work allows us to understand how the traditional view of perception (recovered in phenomenology and present in premodern psychology) as primarily synesthetic is vindicated by the recent understanding of the senses as "modalities" that cross over their functional (*partes extra partes*) determinations: for example, the now demonstrated capacity of human consciousness to have "visual perceptions" through touch, that is possible for blind individuals with the aid of a device that transforms a digital image into electrical impulses on the skin. If perception is something we do, not something that happens to us (like other autonomous internal physiological processes), it is obvious that our intellectual and motor skills are fundamental to cognition.[21] By the same token, the external world—i.e., the city and architecture—*truly matters*, and we do not relate to it as if it were a text in need of interpretation or "information" conveyed to a brain: interpretation comes after we have the world in hand, and in this way architecture affects us along the full range of awareness, from prereflective to reflective. We are "already" in a shared social context and in the "game," rather as we might participate in a sports match, depending on motor intentionality and skills for our perceptions. As Merleau-Ponty points out, the consciousness of the player "is nothing other than the dialectic of milieu and action. Each maneuver undertaken by the player modifies the character of the field and establishes in it new lines of force in which the action in turn unfolds and is accomplished, again altering the phenomenal field."[22] Thompson emphasizes a crucial point for architecture that eluded Heideggerian philosophers like Hubert Dreyfus, was always a difficult question for Merleau-Ponty, and became for poststructuralists a hotly debated issue that denied art its capacity for "meaning as presence": reflective self-awareness is not the only kind of self-awareness. Experience also comprises a prereflective self-awareness *that is not unconscious*. This includes particularly the prereflective bodily self-consciousness profoundly affected by the environment (architecture) that may be passive (involuntary) and intransitive (not object-directed). Thompson adds that there is every reason to think that this sort of prereflective self-awareness animates skillful coping.[23]

Thus—contrary to some fashionable misapplications of the term autopoiesis (a term reserved by Varela and Maturana for metabolic, autonomous life) to parametric architecture and the desire to create "intelligent" buildings that cater to our comfort by emulating the systems of a "computerized mind"—neurophenomenology's understanding of architecture would be as a heteropoietic system, capable of harmoniously complementing the metabolic processes of human consciousness, seeking a balance between the need to provide for a sense of prereflective purposeful action and a reflective understanding of our place in the natural and cultural world. Limits, here, would be articulated not as part of a system (as in a cell) but through language, in view of intersubjective expression. It bears recalling, though this complex issue cannot be developed in a short chapter, that language also has its roots in the prereflective realm of gesture and the body as a primary expressive system. It is not a more or less arbitrary, constructed code. Merleau-Ponty's work is crucial to this issue (as are Heidegger's intuitions): language is "emergent," it "speaks through us" and captures meaning in its mesh; words point toward meanings but never fully coincide with them.[24]

I would argue that the unique gift of architecture is to offer experiences of sense and purpose not in the mere fulfillment of pleasure, but in the *delay* (Duchamp's famous word) that reveals the space of human existence as a space of desire, actually bittersweet, never ending with a punctual homeostasis (i.e., never reduced to the search for ever-increasing comfort or fulfillment). The so-called meaning of existence then appears profoundly grounded in our biology, yet as a true human alternative where desire is never-ending, but may be always sensed as purposeful in our actions amidst appropriate environments, particularly when framed by attuned works of architecture. In other words, architecture's gift is to reveal *the true temporality* of the space of human experience, one that is indeed open to spirituality: the experience of a present moment that, while it can be conceptualized by science (and our clocks) as a quasi-nonexistent point between past and future, is experienced as thick and endowed with dimensions—in a sense, as eternal. This has always been the time "out of time" which is the gift of rituals, festivals, and art, or the time of "silence" evoked by Louis Kahn and Juhani Pallasmaa for architecture. This present "with dimensions" corresponds to Merleau-Ponty's *écart*, the delay between prereflective experience and reflective thought in all its modalities that is paradoxically present in experience and that neuroscience has substantiated.

Indeed, as I have suggested, according to neurophenomenology the formal structure of time-consciousness or phenomenal temporality has an analog in the dynamic structure of neural processes.[25] This uniquely human temporality is generally hidden under scientific and hedonistic interpretations of meaning. Architecture's well-documented gift throughout history, like poetry's, is indeed to allow humans to perceive their sense in the experience of a coincidence of opposites: being and nonbeing beyond theological dogma.[26]

MOOD AND MEANING

Once we start to understand, through recent cognitive science, that our consciousness does not end with our skulls, it becomes easy to grasp that the emotive character of the built environment matters immensely: what matters, in other words, is its material beauty; its power to seduce us on one hand, and its capacity to open up a space of communication for intersubjective encounters on the other. The cognitive sciences' engagement of phenomenology has been productive, and we must expect that in the future this cross-disciplinary pollination will yield important insights for architecture.

Indeed, if the quality of the lived environment is lacking—if we do not even look out to our surroundings for orientation and instead employ technological devices like GPS to find our locations in the world, for instance—our skills are continuously jeopardized and our actions actually reinforce our pathological nihilistic assumptions and the belief that "life is meaningless." Rather than accepting that the built environment is merely a shelter, and all that matters is our possession of a sophisticated computer or an intelligent phone, these insights from neurophenomenology point to the crucial importance of our habitat, one that for humans includes the complexities of material cultures and spoken language. The place of embodied appearance, where we find ourselves through the presence of others, is indeed nothing like the computer screen. Such spaces need to embody appropriate moods or atmospheres to further our spiritual well-being. Architecture has to speak back to us without becoming merely invisible, acting like a numbing drug or like the perfect fit dreamt by functionalism, and today by architects who design increasingly more "intelligent"—i.e., comfortable and efficient—buildings.

In fact, already fed up with functionalism in the mid-twentieth century, Frederick Kiesler imagined in his *Endless House* project an environment that would respond to our moods not by pleasing us (or perhaps simply hiding our mortality) but by challenging us, promoting the use of our imagination, so that every time we turned on the faucet, for example, we would no longer perceive a liquid that circulates, composed of hydrogen and oxygen, but experience instead the real (poetic) nature of water: its qualities as life-giving and primordial liquid, vehicle of purification and remembrance. Such intention offers difficult challenges to a contemporary practice driven by pragmatic and economic imperatives, yet it is a challenge we must take seriously. In other words, sustainability, ecological responsibility, and efficient construction, important as they are, are not enough to fashion a human environment.

Hubert Dreyfus has speculated on the importance of understanding moods for architectural design.[27] It is easy to observe that human actions can change the mood in a room: it can be transformed through a charismatic speaker, lighting effects, artificial acoustics, etc. On the other hand, architects are capable of incorporating in their designed spaces a more lasting mood, one that we may associate with the room itself: solemn, strange, quiet, cheerful, reverential, oppressive, etc. It is important to point out that regardless of these precisions, our architectural experience is always ultimately dependent upon our participation in an event housed in the space; it is in such circumstances that architecture "means."

This contemporary concern is rooted in the romantic concept of *Stimmung*, mentioned above: an attunement that evokes interiority. *Stimmung* is related etymologically to the central questions of harmony and temperance in music, philosophy, and architecture, going back to the origins of European thought in ancient Greece.[28] Significantly, traditional treatises on architectural theory always characterized this concern through the objectivity of mathematics (proportions, geometry), encompassing both form and space. By the end of the eighteenth century in Europe, this understanding was recognized as problematic. In his treatise *Le Génie de l'architecture* (Paris, 1780), Nicolas Le Camus de Mézières addressed the "same" traditional issue but thought that the only way to incorporate the need for harmony in design (an "analogy with our sensations," as he put it) was to characterize the moods or atmospheres of rooms through *words*. He describes a sequence of spaces in a house, rooms with different attributes (light, color, textures, decoration, etc.) related appropriately to the focal actions to which they gave place. It was in this manner that the harmonic potential of architecture—i.e., its meaning—could be sought.[29] Let me emphasize: this expressive potential was set out in words, as descriptive narratives—and no longer in "numbers," as had been traditionally the case when referring to architectural beauty and convenience in most earlier treatises on architecture in the Western tradition—alluding to the "obvious" external harmonies that pervaded the cosmos and man's world.

Indeed, the Cartesian model of reality fails to explain the way moods are normally shared in the everyday world, so at the time when Descartes's dualistic concept became accepted as a fact by the culture at large, architects like Le Camus felt that moods had to be made explicit in language, a vehicle of our primary intersubjectivity, bringing forward what remains a central issue for architectural meaning today. In the everyday world our bodies spontaneously express our moods; others directly pick them up and respond to them. Merleau-Ponty calls this phenomenon intercorporeality: "It is as if the other person's intention inhabited my body and mine his."[30] According to Gaston Bachelard, we literally resonate with another's experience. First there is reverberation, followed by the experience in oneself of resonances, and these eventually have repercussions in the way we perceive the world. This is how the poetic image is communicated, and we can all have the experience of being cocreators.[31]

Now neuroscientists have found an explanation for this important phenomenon in the mirror neurons that fire both when one makes a movement and when one sees another person make that sort of movement: when we observe the actions of others, our nervous system literally "resonates" along with the other.[32] Heidegger had already observed this: "Attunements ... in advance determine our being with one another. It seems as though an attunement is in each case already there, so to speak, like an atmosphere in which we first immerse ourselves ... and which then attunes us through and through."[33] Like an atmosphere, a mood is shared, and is contagious, just like laughter or yawning. This contribution of neuroscience to the understanding of our "virtual" body through mirror neurons has enormous potential to grasp the possibilities of "telepresencing" in multimedia spatial installations, for example,[34] and in the consideration of digital media in design. In all these considerations, however, we must not forget that even more fundamental than neural effects is our embodied consciousness, our intercorporeality; gestures and actions generate habits that are at the root of understanding. We are *primarily* social beings; thus any concern for architectural meaning must build its formal and spatial decisions upon this foundation.[35]

Heidegger specifies further: "Moods are precisely a fundamental manner of being with one another ... and precisely *those* attunements to which we pay no heed at all ... are the most powerful." In a sense, conscious existence, "*Dasein*, is always already attuned. ... There is only ever a change of attunement."[36] Yet being attuned to a situation makes things matter to us: we feel more complete and become participants; our lives matter. This could be the humble yet crucial contribution of architecture in a secular age. But to get there, we must engage language in design practice to articulate human action, avoiding the merely pictorial. Indeed language, particularly in literature, has a greater potential for creating vivid images than "pictures in the mind."[37]

Heidegger recommends spaces that gather self-contained local worlds, gathered around "things thinging." For example, the family meal: a "focal practice" that draws everyone together into a shared mood, so that the action "matters."[38] Such moods "can bring us in touch with a power that we cannot control and that calls forth and rewards our efforts," a power that could be recognized as sacred. The sense that the mood is shared is constitutive of the excitement, as used to happen in traditional rituals and still does in some contemporary performances, or in our experience of art. The architect can therefore try to bring about the appropriate moods for human actions that reveal life as purposeful by designing spaces that are open to an appropriate range of moods. I would argue that literary language can describe these possibilities as one imagines a proposed space being used, in manifold contexts, to invite in the unexpected: thus architecture is never static, neutral, or merely devoted to one use.

PEDAGOGICAL COROLLARY

Acknowledging the complexities of an embodied and situated human consciousness, as sketched above, should have important consequences for architecture in the twenty-first century. Despite the genealogy I have outlined, many such consequences are only now starting to be understood. For the moment, these remarks should begin to suggest revisions to long-held beliefs that characterize architectural education. We are now in a position to understand that design cannot be dictated by functions, algorithms, or any sort of compositional *mathesis*, for the issues of architecture are never simply technological or aesthetic. In fact, the current emphasis on novel forms, simply for the sake of novel effects with no regard for materiality, seems hardly justifiable. While there is nothing that may appear in human experience to be totally "meaningless," usually such products are obviously no more than expensive thrills. Indeed, the question is not merely to create fantasies, whether biomorphic or biomimetic, cultivating novelty for its own sake. On the other hand, architectural design is not problem-solving, in the way that one might resolve an equation that encompasses all its variables. Even if architectural practice may treat its products as mere services or commodities, schools should resist these assumptions, and emphasize the ethical and poetic dimensions of our discipline. The new central concern should be to prepare the future architect to use her imagination to *make* poetic artifacts and spaces with character, resonant with the human situations they house, engaging dimensions of consciousness that are in fact usually stifled by conventional educational paradigms, rather than simply *planning* efficient buildings.

Designing a poetic architecture is not a merely intuitive operation, or an unreflective action, but rather the continuation of a practical philosophy in the tradition of Aristotle's *phronesis*, a kind of rational wisdom capable of considering actions that can deliver desired effects. This is not simply a skill or *technē*, however, since it involves not only the ability to decide how to achieve a certain end, but also the ability to reflect upon and determine good ends consistent with the aim of living well: a wisdom profoundly grounded in culture and historical understanding, and also associated with political ability.

Indeed, teaching the future architect the elements of *praxis*—how to articulate an appropriate and ethical position that becomes incorporated in the project—is paramount in view of this intertwining of embodied consciousness and world. Instrumental methodologies that seem to cancel the distance between theory (applied science) and practice are a dangerous contemporary delusion. We must trust the words of our spoken languages that have their roots in the same flesh of the world as our actions, despite their "opacity" when it comes to determinate meanings. Thus, the teaching of a "philosophical history"—understanding architecture as one of the humanities—is fundamental: sharing stories capable of making the artifacts and buildings from our traditions reveal their full values through a hermeneutic process. In the spirit of Friedrich Nietzsche and Hannah Arendt, this is a "history for the future," one that is meant to enhance our vitality and creativity rather than one that may immobilize us through useless data or unattainable models.

When they are understood in the historical contexts (scientific, religious, philosophical or mythical) of their makers, the architecture and the words which have articulated the *praxis* of other times and places will be revealed as both emotionally moving and discursively pertinent, capable of orienting the inhabitant and providing poetic modes of dwelling. This process of interpretation, appropriating through our own questions that which is acknowledged as truly distant, makes it possible to render the voices of our architectural heritage into our own specific time and political and social contexts, providing the best possible education. Thus the experience of beauty may truly enhance our creativity, avoiding either mere copying (conservative revivalisms) or irrelevant formal novelty. In other words, we should teach history to discover the hidden treasures in our traditions, which, as Giorgio Agamben has stated, we can now—paradoxically, in view of the well-publicized disregard for the past of our future-driven technological society—truly start to articulate fully, as we grasp the falsity of a single historical narrative. Thus the future architect may learn to make responsible promises, engaging the fictional character of the discipline and imagining, in words, possible ways of addressing the political reality, proposing novel programs to attain our humanness and shaping form toward its expressive intention.

Finally, I would like to touch upon the unavoidable issue of digital tools and their place in design education. I would insist on the importance of keeping a critical yet open mind. Their limitations have often been pointed out, since architects mostly use software that takes for granted the reality of Cartesian space, and generally assumes that lived experience can be reduced to numerical data. Indeed, dedicated software like CAD and BIM/Revit, which have built-in Cartesian coordinates that manifest as the space of design, are thus inherently misleading. This reduction of lived tactile and qualitative place to geometric space is the most dangerous fallacy for a future (and practicing) architect; it leads to the belief that architectural meaning lies simply in the novel manipulation of forms. Nevertheless, it must be stressed that this assumption, though made more dangerous by the powers inherent in tools (now related to robotic fabrication), was not the result of the so-called "digital revolution": it was already present in the (analog) mechanisms of architectural representation rooted in descriptive geometry at the inception of modern (university-based) architectural education, first made explicit in the early-nineteenth-century theories of Durand.[39]

Teaching must avoid the delusions created by these tools, and focus on developing multiple skills which truly enrich the future architect's capacity to perceive the qualities of the life world, engaging modes of representation that may render such qualities with the specific tasks at hand, acknowledging that materiality may have priority over form for emotional embodied meanings, a condition which is particularly evident when form is used to describe a building as an object, implicitly decontextualized and dematerialized and fetishized for its complex geometries. Indeed, we should teach the perception of emotional and meaningful space through artifacts that may be accessible in the school environment (like painting, literary fiction, film, etc.), emphasizing the possibility of translation, for in order to teach the poetic utterance, the point of departure must be poetic: beauty breeds and inspires beauty through immediate, emotionally charged perception. In other words, the objective should be an opening awareness of possible realities, the development of critical approaches to the "tools" of the architect. In school, rather than assuming that "representations" are neutral, the products of the process must be *fully* valorized. It is possible to teach awareness of the wonders that we can reveal, to ourselves, through human work. This necessitates patience and open-endedness, cultivating the "thick present" that characterizes embodied perception and that neuroscience now has called to our attention, understanding that tools and skills modify perception itself. Every moment of a search is liable to turn up poetic disclosures which may eventually be translated into different dimensions, and thus start to constitute the future architect's own vision.

The skills associated with hand drawing are simply different from those involved in manipulating a mouse. And yet it is interesting to consider how neuroscience has "demonstrated" the reality of the virtual body, one that can feel threatened viscerally at a distance when others do it violence—a well-documented condition in Paul Sermon's 1994 installation *Telematic Dreaming*, performed by Susan Kozel.[40] This is an important and complex question.[41] There are now attempts to use software that questions its Cartesian ordering system to allow designing through a more embodied engagement with the process—as, for example, in "3DS Max," in which cinematic vision is simulated by a virtual camera, which the designer can deploy to experience space from a "first-person" vantage point. Of course, cognitive science demonstrates that our vision is never merely homologous to photographic perspective, even in motion, so the analogy is limited. More recent software allows one to work in non-Cartesian space and draw without having to use precise measurements, facilitating drawing as a thinking process. Such is the case in software like "Mudbox," originally created to design artifacts based primarily in the service of narrative structures, such as interactive "3D" applications, video games, animated film, and visual effects. As Hubert Dreyfus concludes in his analysis of virtual architecture in "Second Life," though, while an avatar seems to pick up on the moods of others in interaction, the moods of rooms are not communicated to avatars in the same way as they are immediately perceived by our living, embodied consciousness.[42] If one agrees that the design of appropriate, harmonious "atmospheres" that recover the qualitative dimensions of place while making space for habitual worlds of skilled engagement is of the essence in architecture, this is one of the crucial questions opened up by neuroscience: the reality of one's own body as a phantom, "constructed temporarily by the human brain for convenience," as explained by V. S. Ramachandran,[43] must be a central consideration in the reassessment of digital tools for design both in practice and in education.

NOTES

This chapter is an elaboration of my responses to questions posed by Sarah Robinson during the panel discussion of the "Minding Design" symposium, November 9, 2012.

1. See Alberto Pérez-Gómez, *Architecture and the Crisis of Modern Science* (Cambridge, MA: MIT Press, 1983).

2. See Claude Perrault, *Ordonnance for the Five Kinds of Columns after the Method of the Ancients*, trans. Indra Kagis McEwen from the 1683 first edition, with my own introductory study (Santa Monica: Getty Center, 1993); and Claude Perrault, *Les dix livres d'architecture de Vitruve* (Paris, 1684).

3. Human visual perception includes peripheral vision and haptic qualities; it is not passive; our understanding of depth is the result of our physical motor engagement with the world. Visual perception is therefore not analogous to a constructed perspective. This problem been discussed exhaustively by Maurice Merleau-Ponty, *Phenomenology of Perception*, trans. D. Landes (New York: Routledge, 2012); and in his essays on art, particularly "Eye and Mind" and "Cezanne's Doubt," collected in G. Johnson, ed., *The Merleau-Ponty Aesthetics Reader* (Evanston: Northwestern University Press, 1993). See also Alva Noë, *Action in Perception* (Cambridge, MA: MIT Press, 2004). The far-reaching implications of this issue for architectural representation have been discussed in Alberto Pérez-Gómez and Louise Pelletier, *Architectural Representation and the Perspective Hinge* (Cambridge, MA: MIT Press, 1997).

4. See Georges Gusdorf, *Fondements du savoir romantique* (Paris: Payot, 1982). Gusdorf has been instrumental in describing the importance of this revolution and its connections to later Continental philosophy and phenomenology.

5. The main figures are Friedrich Schelling, Friedrich Schlegel, Novalis (Georg Friedrich Freiherr von Hardenberg), and Carl Jacobi.

6. See Merleau-Ponty, *Phenomenology of Perception*; and especially M. C. Dillon, ed., *Écart et Différence: Merleau-Ponty and Derrida on Seeing and Writing* (Atlantic Highlands, NJ: Humanities Press, 1997), chs. 1, 6, 11.

7. Friedrich Wilhelm Joseph Schelling, *Lettre sur le dogmatisme et le criticisme*, trans. S. Jankelevitch (Paris: Aubier, 1950), 109; emphasis added.

8. See Daniel Heller-Roazen, *The Inner Touch: Archaeology of a Sensation* (New York: Zone Books, 2007).

9. See Leo Spitzer, *Classical and Christian Ideas of World Harmony: Prolegomena to the Interpretation of the Word "Stimmung"* (Baltimore: Johns Hopkins University Press, 1963).

10. See Francisco Varela, Evan Thompson, and Eleanor Rosch, *The Embodied Mind* (Cambridge, MA: MIT Press, 1991), especially chs. 10 and 11.

11. This argument has been brilliantly put forward by Hans-Georg Gadamer, *Reason in the Age of Science* (Cambridge, MA: MIT Press, 1981), ch. 8.

12. Friedrich Nietzsche, "On the Advantages and Disadvantages of History for Man," in *Untimely Meditations* (Cambridge: Cambridge University Press, 1983).

13. Evan Thompson, *Mind in Life: Biology, Phenomenology, and the Sciences of the Mind* (Cambridge, MA: Harvard University Press, 2007), 4.

14. Ibid., 5.

15. Ibid., 8.

16. Ibid., 9.

17. Ibid., 10.

18. See Antonio Damasio, *Descartes' Error* (Toronto: Penguin Books, 2005); and *Looking for Spinoza: Joy, Sorrow, and the Feeling Brain* (Toronto: Harcourt, 2003).

19. Thompson, *Mind in Life*, 15; emphasis added.

20. See, for example, Shaun Gallagher, *How the Body Shapes the Mind* (Oxford: Clarendon Press, 2006); and Louise Barret, *Beyond the Brain: How Body and Environment Shape Animal and Human Minds* (Princeton: Princeton University Press, 2011). While not all on exactly the same footing, these books contribute enormously to our understanding of the issues raised by enactive cognition and neurophenomenology.

21. Alva Noë, *Out of Our Heads: Why You Are Not Your Brain and Other Lessons from the Biology of Consciousness* (New York: Hill and Wang, 2009), 7. See also Noë's more technical *Action in Perception*.

22. *Maurice Merleau-Ponty Basic Reader*, ed. Thomas Baldwin (London: Routledge, 2013), 53.

23. Thompson, *Mind in Life*, 315–316.

24. See Maurice Merleau-Ponty, "The Phenomenology of Language," in *Signs* (Evanston: Northwestern University Press, 1964), 84–97.

25. Thompson, *Mind in Life*, 356–357.

26. This is also Octavio Paz's universal definition for a "poetic image." Octavio Paz, *The Bow and the Lyre* (Austin: University of Texas Press, 1991).

27. Hubert Dreyfus, "Why the Mood *in* a Room and the Mood *of* a Room Should Be Important to Architects," in *From the Things Themselves: Architecture and Phenomenology* (Kyoto: Kyoto University Press, 2012), 23–39.

28. I examine this problem extensively in a work in progress, *Attunement: In Search of Architectural Meaning*.

29. Nicolas Le Camus de Mézières, *The Genius of Architecture; or the Analogy of That Art with Our Sensations*, trans. D. Britt (Santa Monica: Getty Center, 1992).

30. Dreyfus, citing Merleau-Ponty, "Why the Mood *in* a Room," 26.

31. Cited by Susan Kozel, *Closer: Performance, Technologies, Phenomenology* (Cambridge, MA: MIT Press, 2007), 25.

32. This is the neurological phenomenon that now explains the "phantom limb" syndrome of amputees. See V. S. Ramachandran and Sandra Blakeslee, *Phantoms in the Brain: Probing the Mysteries of the Human Mind* (New York: William Morrow, 1998).

33. Dreyfus, citing Heidegger, "Why the Mood *in* a Room," 27.

34. See in this respect the fascinating work of Susan Kozel and her reflections on her work through phenomenology, in *Closer*.

35. See Nick Crossley, *The Social Body: Habit, Identity, Desire* (London: Sage, 2001).

36. Martin Heidegger, *Fundamental Concepts of Metaphysics* (Bloomington: Indiana University Press, 1995), 67–68.

37. This is the main argument developed by Elaine Scarry, *Dreaming by the Book* (Princeton: Princeton University Press, 2001), 3–9.

38. Dreyfus, citing Heidegger, "Why the Mood *in* a Room," 35.

39. See Pérez-Gómez, *Architecture and the Crisis of Modern Science*, chs. 8, 9.

40. Kozel, *Closer*, 88–90, 92–101.

41. See Alberto Pérez-Gómez and Angeliki Sioli, "Drawing *with/in* and Drawing *out*, a Redefinition of Architectural Drawing through Edward Casey's Meditations on Mapping," *Festschrift in Honor of Edward Casey* (in preparation).

42. Dreyfus, "Why the Mood *in* a Room," 32–34.

43. Ramachandran and Blakeslee, *Phantoms in the Brain*, 58.

CONTRIBUTORS

Thomas D. Albright is the Director of the Vision Center Laboratory and Conrad T. Prebys Chair in Vision Research at the Salk Institute of Biological Sciences. He is the current president of the Academy of Neuroscience for Architecture.

Michael Arbib is the Director of the USC Brain Project, the Fletcher Jones Professor of Computer Science, and a Professor of Biological Sciences, Biomedical Engineering, Electrical Engineering, Neuroscience and Psychology at USC. His fortieth book is *How the Brain Got Language*. He is a member of the Board of Trustees of the Academy of Neuroscience for Architecture.

John Paul Eberhard is the founding president of the Academy of Neuroscience for Architecture and the second recipient of the AIA College of Fellows Latrobe Fellowship for Research. He is the author of *Brain Landscape: The Coexistence of Neuroscience and Architecture* and *Architecture and the Brain: A New Knowledge Base from Neuroscience.*

Melissa Farling, FAIA, is a Research/Project Director at Jones Studio, Co-Chair for the AIA's Academy of Justice for Architecture Research Committee, and a member of the Academy of Neuroscience for

Architecture Advisory Council. For twenty-five years she has focused on criminal justice facilities and large-scale public projects, and was one of the principal investigators on a National Institute of Corrections-funded study to examine impacts of views of nature on stress levels. She is a contributing author to the AIA *AAJ Sustainable 2030: Green Guide to Justice* and the *Arizona School Design Primer: The Basic Elements of School Design.*

Vittorio Gallese is a Professor of Neurology in the Department of Neuroscience, University of Parma. He was a member of the original team of four scientists who discovered mirror neurons in the early 1990s. He has authored over 150 papers and numerous books on intersubjectivity, language, social cognition, perception and cognition, psychiatry, phenomenology, and aesthetics. He currently directs two teams of researchers in the areas of neurophysiology and cognitive neuroscience.

Alessandro Gattara, an architect who has practiced in New York and Italy, is pursuing his Ph.D. in architecture with Vittorio Gallese at the University of Parma. A graduate of the Politecnico di Milano and the Bocconi Institute and the former editor of the journal *SdS*, he is a contributor to the magazine *Area* and the web site *Gizmo*. He is the Italian translator of Harry Mallgrave's *Architecture and Embodiment.*

Aris Georges, this book's graphic designer, is professor of Architecture at Taliesin, the Frank Lloyd Wright School of Architecture, a graphic designer, and an artist. He studied architecture at the University of Florence and at Taliesin. He is the author and graphic designer of *Nature Patterns*, forthcoming from Routledge.

Mark L. Johnson is Knight Professor of Liberal Arts and Sciences in the Department of Philosophy at the University of Oregon. Well known for his contributions to embodied philosophy, cognitive science, and cognitive linguistics, he is the author of numerous books and articles including *Metaphors We Live By*, with George Lakoff, and *The Meaning of the Body.*

Harry Francis Mallgrave, an honorary fellow of RIBA, is a professor of architecture at Illinois Institute of Technology. For nearly two decades he served as the Editor of Architecture and Aesthetics for the "Text and Documents Series" of the Getty Research Institute. He is the author of numerous books and articles, including the groundbreaking *The Architect's Brain: Neuroscience, Creativity and Architecture* and *Architecture and Embodiment: The Implications of the New Sciences and Humanities for Design.*

Iain McGilchrist is a former Fellow of All Souls College, Oxford, a Fellow of the Royal College of Psychiatrists, and former Consultant Psychiatrist and Clinical Director at the Bethlem Royal & Maudsley Hospital, London. He is the author of numerous articles, papers, and books on diverse subjects, including *Against Criticism*. His most recent book is *The Master and His Emissary: The Divided Brain and the Making of the Western World.*

Juhani Pallasmaa, architectural theorist and practitioner, is an Honorary Member of SAFA Society of Finnish Architects, AIA American Institute of Architects, and RIBA Royal British Institute of Architects, and he has received numerous Honorary Doctorates in Architecture, Technology and the Arts. The former director of both the Finnish Museum of Architecture and the Department of Architecture at Helsinki University of Technology, he has authored over 30 books including *The Eyes of the Skin: Architecture and the Senses, The Thinking Hand*, and *The Embodied Image.*

Alberto Pérez-Gómez received his undergraduate degree in architecture and engineering in Mexico City, did postgraduate work at Cornell University, and completed his M.A. and Ph.D. at the University of Essex in England. The Saidye Rosner Bronfman Professor of Architectural History at McGill University, he is the author of numerous books and articles including *Architecture and the Crisis of Modern Science* and *Built upon Love*.

Sarah Robinson is a practicing architect who studied philosophy at the University of Wisconsin-Madison and the University of Fribourg in Switzerland before attending the Frank Lloyd Wright School of Architecture, where she earned her M.Arch and later served as the founding chair of the Board of Governors. She is the author of *Nesting: Body, Dwelling, Mind*, and lives in Pavia, Italy.

FIGURE CREDITS

1.1 Photograph by Harry Mallgrave.

1.2 Photograph by Harry Mallgrave.

1.3 From Gottfried Semper, *Der Stil in den technischen und tektonischen Künsten, oder praktische Aesthetik* (Frankfurt am Main: Verlag für Kunst und Wissenschaft, 1860–1863).

1.4 From M. E. Sadler, ed., *The Eurhythmics of Jaques-Dalcroze* (Boston: Small Maynard and Company, 1913).

1.5 From M. E. Sadler, ed., *The Eurhythmics of Jaques-Dalcroze* (Boston: Small Maynard and Company, 1913).

1.6 Photograph courtesy of Sarah Robinson.

1.7 Photograph courtesy of Georges Jansoone, Creative Commons Attribution-Share 3.0 Unported.

1.8 Photograph by Harry Mallgrave.

1.9 Photograph by Harry Mallgrave.

1.10 From Gottfried Semper, *Der Stil in den technischen und tektonischen Künsten, oder praktische Aesthetik* (Frankfurt am Main: Verlag für Kunst und Wissenschaft, 1860–1863).

3.1 Photograph by Don Mammoser, Shutterstock.com.

3.2 Courtesy of the Aulis Blomstedt Estate, Helsinki.

3.3 Courtesy of Juhana Blomstedt.

3.4 Courtesy of Harumi Klossowska de Rola.

3.5 Courtesy of the Frank Lloyd Wright Foundation.

3.6 Courtesy of the Frank Lloyd Wright Foundation.

3.7 Courtesy of Artists Rights Society (ARS), New York/SIAE, Rome. © 2014.

3.8 My Good Images/Shutterstock.com.

3.9 Photograph courtesy of Rauno Träskelin.

3.10 Creative Commons.

3.11 Courtesy Turid Hölldobler-Forsyth; in Karl von Frisch and Otto von Frisch, *Animal Architecture* (New York: Harcourt Brace Jovanovich, 1974).

3.12 Courtesy of Artists Rights Society (ARS), New York/SIAE, Rome. © 2014.

3.13 Photograph courtesy of Iwan Baan.

4.3 From J. Szentágothai and M. A. Arbib, *Conceptual Models of Neural Organization* (Cambridge, MA: MIT Press, 1975).

4.4 Courtesy of Marc Jeannerod and Jean Biguer.

4.5 Courtesy of Artists Rights Society (ARS), New York. © Herscovici 2014.

4.6 Photograph courtesy of Enrico Fegu.

4.7 Photograph courtesy of Anna Smith.

4.8 Photograph courtesy of Anna Smith.

5.1 Courtesy of Iain McGilchrist.

5.2 Public domain.

5.3 Public domain.

5.4 Public domain.

5.5 Creative Commons, Zairon 2007.

5.6 Photograph courtesy of the Western Pennsylvania Conservancy.

6.1 sarra22/Shutterstock.com.

6.2 cdrin/Shutterstock.com.

6.3 Public domain.

6.4 Courtesy of John Paul Eberhard.

6.5 Courtesy of John Paul Eberhard.

6.6 Courtesy of John Paul Eberhard.

6.7 Courtesy of John Paul Eberhard.

7.1 Courtesy of the Hundertwasser Foundation.

7.2 Photograph courtesy of David Sundberg.

7.3 Photographs courtesy of Sarah Robinson.

7.4 Photograph courtesy of the Western Pennsylvania Conservancy.

7.5 Photograph courtesy of Hedrich Blessing.

7.6 Photograph courtesy of Hedrich Blessing.

8.1 Photograph courtesy of Sarah Robinson.

8.2 Photograph courtesy of Creative Commons, Marie-Lan Nguyen 2009.

8.3 Courtesy of Artists Rights Society (ARS), New York. © 2014.

8.4 Courtesy of Artists Rights Society (ARS), New York. © 2014.

8.5 Courtesy of Artists Rights Society (ARS), New York. © 2014.

8.6 Photo courtesy of Lombardini 22, DEGW Milano.

9.1 Photograph courtesy of A. F. Payne.

9.2 Photograph courtesy of Matthew Salenger.

9.3 Photograph courtesy of Brian Farling.

9.4 Photograph courtesy of Bill Timmerman.

10.1 Courtesy of the Journal of Physiology, Wiley and Sons, London.

10.2 Photograph by Alex Sun/Shutterstock.com.

10.3 Photograph by Komkrit Preechachanwate/Shutterstock.com.

10.4 Photograph by Sarah Robinson.

10.5 Photograph by Lazio/Shutterstock.com.

10.6 Photo by Doug Reed.

10.7 Photo by Sarah Robinson.

10.8 Courtesy of the Frank Lloyd Wright Foundation.

10.9 Courtesy of the Frank Lloyd Wright Foundation.

10.10 Photograph by Vlacheslav Lopatin/Shutterstock.com.

10.11 Photograph by Mary Mueller.

INDEX

"f" refers to figures.